VIEWING NORFOLK SOUTHERN RAILWAY

On September 23, 2022, an eastbound Norfolk Southern Railway (NS) freight train is passing through the Fostoria, Ohio, "Iron Triangle" (three mainline railroad tracks cross each other forming a 5.6-acre triangular-shaped rail park) headed by 4,400-horsepower NS locomotive No. 4179 (that was originally built by General Electric [GE] as type Dash 9-44CW locomotive No. 8965 and rebuilt as type AC44C6M in March 2019 by Wabtec/GE at Erie, Pennsylvania, and renumbered 4179). (*Kenneth C. Springirth photograph*)

Distributed Power Unit (DPU) Kansas City Southern Railway 4,400-horsepower GE type ES44AC (Evolution Series) locomotive No. 4484 is in the middle of the above NS train at Fostoria, Ohio, on September 23, 2022. A DPU (controlled by the leading locomotive) is a remotely controlled locomotive that can be at the middle or end of a train to allow pulling a longer train with more horsepower. (*Kenneth C. Springirth photograph*)

VIEWING NORFOLK SOUTHERN RAILWAY

Beth Anne Keates and Kenneth C. Springirth

AMERICA THROUGH TIME®
ADDING COLOR TO AMERICAN HISTORY

In July 2022, Norfolk Southern Railway (NS) 4,300-horsepower six-axle diesel locomotive No. 1111 (built by the Electro Motive Division of General Motors Corporation [EMD] type SD70ACe placed in service on March 13, 2014) is trailing on a westbound intermodal train through the borough of Wyomissing in Berks County, Pennsylvania. (*Brian A. Keates photograph*)

On the top portion of cover: On the top portion of front cover: On March 19, 2023, Norfolk Southern Railway (NS) 4,400-horsepower type ES44AC locomotive No. 8102 built in February 2012 by General Electric Company (GE) is westbound powering an intermodal train through Altoona, Pennsylvania. This locomotive was painted in the Pennsylvania Railroad heritage paint scheme at Juniata, Pennsylvania, on April 18, 2012. (*Brian A. Keates photograph*)

On the bottom portion of cover: Norfolk Southern Railway (NS) locomotive No. 4003 is heading an eastbound train at Macungie, Pennsylvania, on May 20, 2023. This locomotive was originally NS No. 8839 GE type Dash 9-40C built in February 1995 and rebuilt as type AC44C6M by the NS Roanoke Locomotive Shop on October 19, 2016, painted in special black, Tuscan red, gray, and white scheme. (*Brian A. Keates photograph*)

On the back cover: NS locomotive No. 1800 is passing through Wyomissing, Pennsylvania, on August 21, 2022. This locomotive was originally NS No. 2537 type SD70 built by Electro Motive Division of General Motors Corporation (EMD) in September 1994. It was rebuilt into type SD70ACC by Progress Rail at Muncie, Indiana, in March 2018 and painted in a yellow, gray, black, and white scheme to highlight the conversion of the locomotive from DC to AC traction. (*Brian A. Keates photograph*)

Fonthill Media Inc.
www.fonthillmedia.com
office@fonthillmedia.com

First published 2023

ISBN 978-1-63499-492-7

Typeset in Mrs Eaves XL Serif Narrow
Printed and bound in England

Contents

Acknowledgments

Thanks to the Erie County (Pennsylvania) Public Library's Heritage Room in the Blasco Memorial Library for their extensive local history file and excellent staff, plus the excellent staff in the Iroquois Branch Library. The *Official Guide of the Railways December 1962* and *Travelers Official Guide of the Railways June 1870* were used for passenger train schedule information. Robert A. Elder, Beth Anne Keates, Brian A. Keates, Virginia M. Springirth and Kenneth C. Springirth provided pictures. A number of pictures came from postcards and were (*Used with permission from Audio-Visual Designs [www.audiovisualdesigns.com]*). The names of the photographers who took the pictures for those photographs are as follows: Howard W. Ameling, Stephen D. Bogen, Marvin H. Cohen, Bob Collins, Richard J. Cook, Dan Cupper, David L. Foster, Jim Fritts, Jim Hawk, Joe Kmetz, Mac Owen, Max Robin, William Rosenberg, Jim Shaughnessy, Carl H. Sturner, William Volkmer, R. R. Wallin, Don Wood, and Douglas C. Wornom.

This book is dedicated to Jacob A. Keates and Andrew L. Keates, the sons of Beth Anne and Brian A. Keates.

This book is also dedicated to Kathleen May Ruggio, the daughter of Virginia and Kenneth C. Springirth.

A westbound NS freight train headed by 4,400-horsepower six-axle diesel locomotive No. 9932 (type D9-44CW built by GE and placed in service on November 1, 2004) is passing by the 5.6-acre Fostoria, Ohio, Iron Triangle Rail Park where three railroad lines meet (two CSX and one NS) on September 23, 2022. Open 24/7 every day of the year, the Fostoria Rail Preservation Society maintains the viewing area and rest room facilities. (*Kenneth C. Springirth photograph*)

Introduction

The Norfolk & Western Railroad and Southern Railway came together on June 1, 1982, forming the Norfolk Southern Railway which is a subsidiary of the Norfolk Southern Corporation. Operating in twenty-two states and the District of Columbia, the Norfolk Southern Railway (NS) handles industrial, agricultural, forest, chemical, consumer, metal, construction products, automobiles, plus automobile parts and coal. While intermodal containers and trailers have become the most common commodity carried by NS, transporting of coal declined from 157.5 million tons in 2009 to 73 million tons in 2021 as noted by NS Annual Reports.

An important financial measure for a railroad is the operating ratio which is the total operating expense of a company divided by the company's net sales. According to NS Annual reports, the operating ratio declined from 75.4 percent in 2009 to 65.4 percent in 2018. To continue that trend, NS introduced Precision Scheduled Railroading (PSR) under TOP21 which stands for Thoroughbred Operating Plan for 2021. PSR strives to minimize car dwell time in yards, minimize car classifications, and run longer and fewer trains to reduce operating costs. In PSR, adherence to train schedules is important because one train's inbound power is often the locomotive for the next outbound train. The NS 2021 Annual Report showed the operating ratio was reduced from 64.7 percent in 2019 to 60.1 percent in 2021. NS railway operating revenues increased 13.82 percent from $9,789 million in 2020 to $11,142 million in 2021. NS net income increased 26.5 percent from $3,486 million in 2020 to $4,447 million in 2021. Using data from NS Annual Reports, the average number of NS employees per year declined from 30,456 in 2015 to 18,500 in 2021 which was a decline of 11,956 or 39.26 percent. However, the 2021 NS Annual report noted: "While our focus on productivity and revenue growth produced record financial results, our service impacted by labor shortages across key areas of our network-was not where it needed to be. Accordingly, we have increased our hiring and retention efforts, and begun development of a new operating plan that is fully aligned around a return to the service levels our customers expect." The goal of this book is to focus on NS and its heritage.

Painted in the special Operation Lifesaver paint scheme, NS 3,000-horsepower four-axle EMD type GP59E diesel locomotive No. 4655 is switching out cars on a local train at the borough of Sinking Spring in Berks County, Pennsylvania, in April 2021. This was rebuilt on January 30, 2014 from NS EMD GP59 locomotive No. 4637 that was originally built in December 1989. (*Brian A. Keates photograph*)

Map of Norfolk Southern Raiway

(Not to scale)

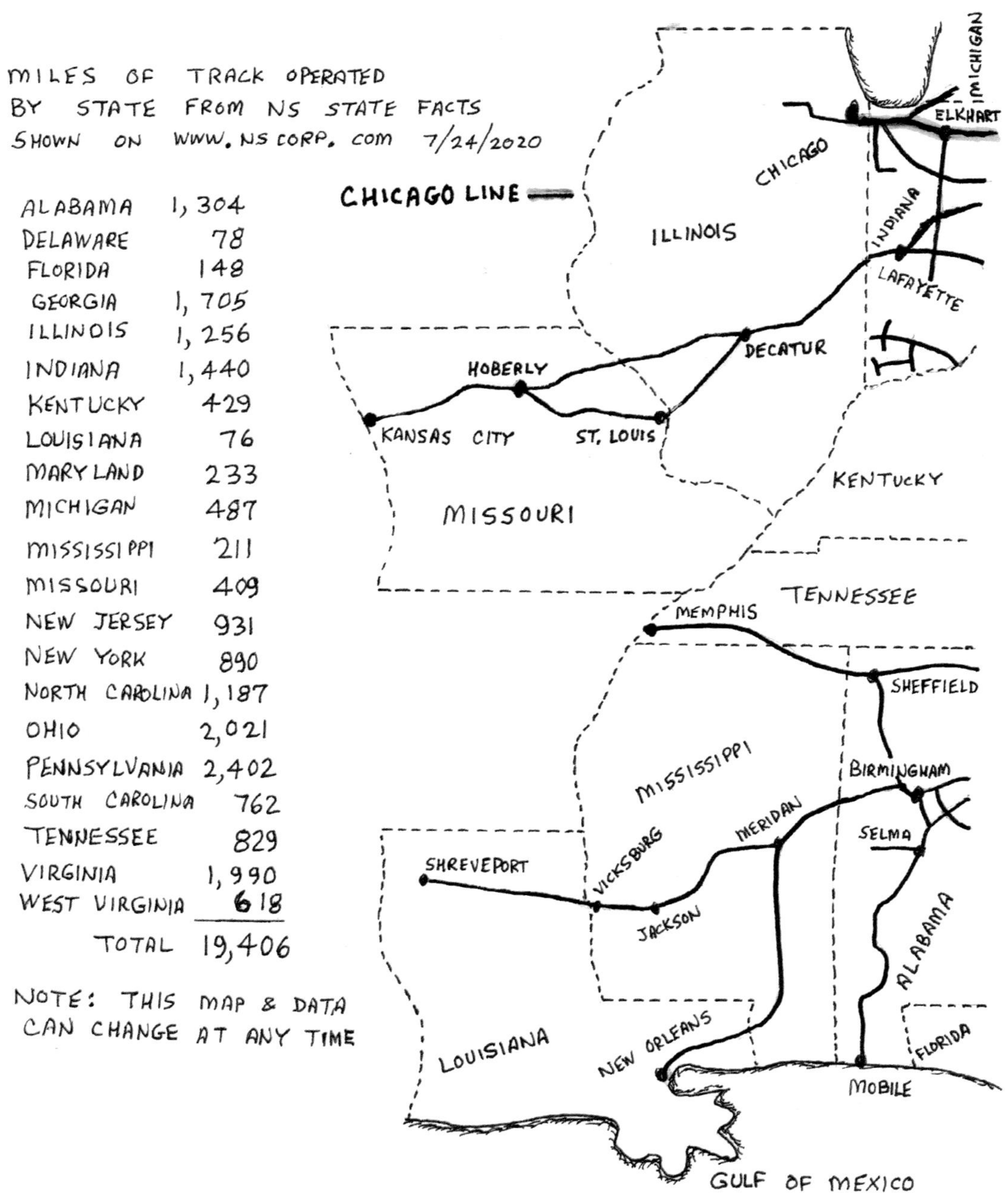

K. C. Springirth 7/24/2020

The map of the Norfolk Southern Railway is based on the 2020 fact sheets for each state found on the Norfolk Southern Railway www.nscorp.com website.

Map of Norfolk Southern Raiway

(Not to scale)

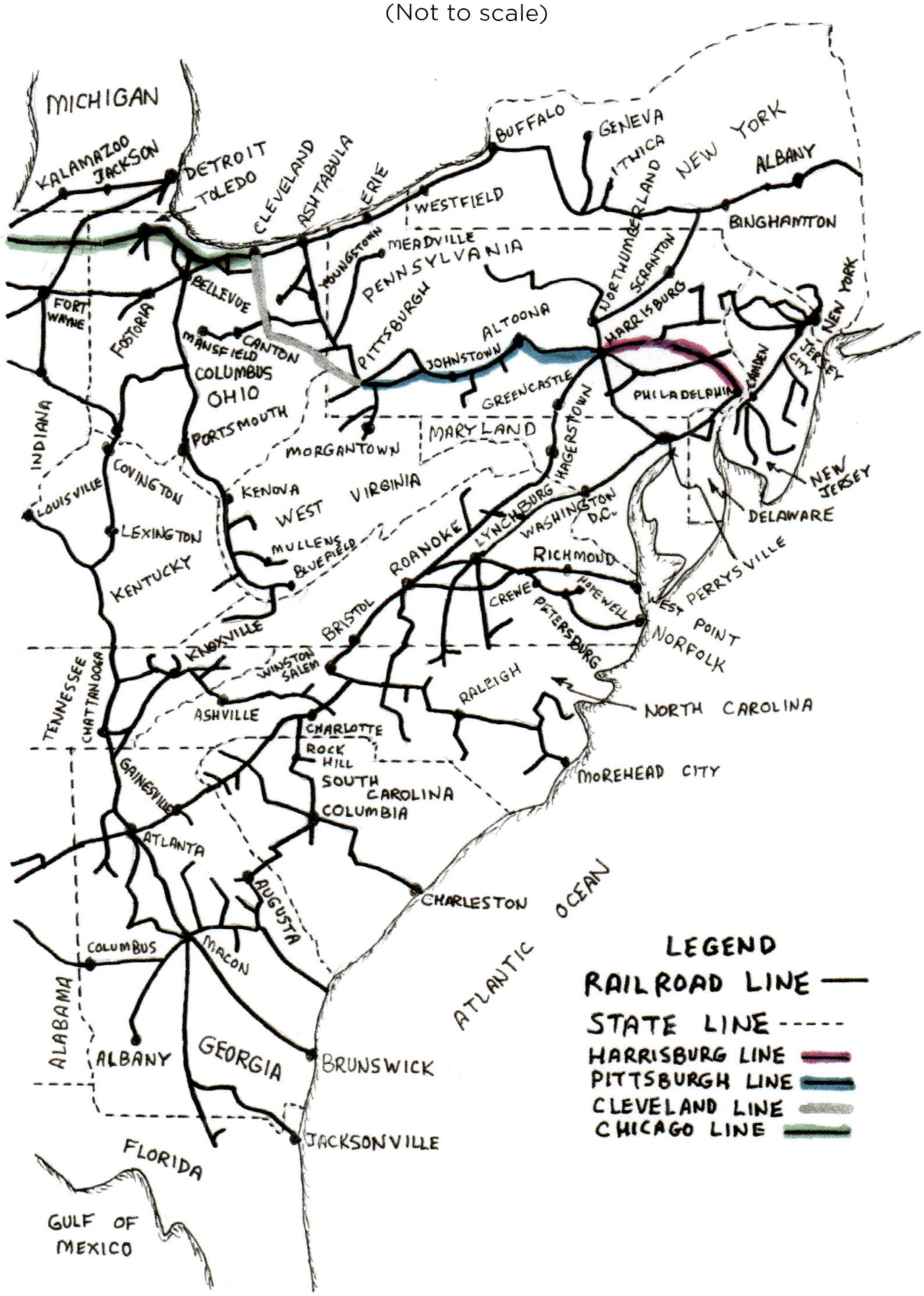

K. C. Springirth 7/24/2020

The above map is based on the 2020 fact sheets for each state found on the Norfolk Southern Railway www.nscorp.com website. Formed from two former Reading lines, the day Conrail began operations on April 1, 1976, the Harrisburg Line is shown as one line from Philadelphia west to Harrisburg.

On a snow-covered February 15, 2021, westbound six-axle 4,400-horsepower Evolution Series Tier 4 NS locomotive No. 3639 (GE type ET44AC placed in service on June 15, 2016) is leading a general manifest train over track one at Gallitzin, Pennsylvania. (*Brian A. Keates photograph*)

Westbound NS train headed by Burlington Northern Santa Fe Railway six-axle 4,400-horsepower locomotive No. 4244 (GE type ES44C4 built in 2016) and trailing 4,400-horsepower NS locomotive No. 7659 (GE type ES44AC placed in service on February 27, 2006) are crossing Walbridge Road in Harborcreek Township, Erie County, Pennsylvania, on October 10, 2022. (*Kenneth C. Springirth photograph*)

1

Predecessor Railroads

Norfolk Southern Railway

The Elizabeth City & Norfolk Railroad was formed on January 20, 1870, and by 1880, the line opened from Berkley, Virginia, to Edenton, North Carolina, with the name changed to Norfolk Southern Railroad on February 1, 1883. After receivership in 1889, it reorganized in May 1891 as the Norfolk & Southern Railroad. On January 21, 1942, the railroad reorganized as the Norfolk Southern Railway. It was purchased by the Southern Railway on January 1, 1974 and merged into the Carolina & Northwestern Railway. Norfolk Southern Railway Corporation was created on June 1, 1982 as a holding company to acquire Norfolk Southern Railway and Southern Railway. It was headquartered in Norfolk, Virginia, until 2021 when its headquarters was moved to Atlanta, Georgia.

Pennsylvania Railroad

The first part of what later became the Pennsylvania Railroad (PRR) was the Philadelphia and Columbia Railroad which opened on October 7, 1834. Following the opening of the Horseshoe Curve on February 15, 1854, on January 1, 1856, the Harrisburg to Pittsburgh line was completed. During 1866, the PRR began building its own locomotives at its Altoona shops. On July 1, 1869, the PRR leased the Pittsburgh, Fort Wayne & Chicago Railway (PFW&C) which gave it a direct route to Chicago. However, on April 1, 1871, the PFW&C was transferred to the newly formed Pennsylvania Company with operation transferred back to the Pennsylvania Railroad on January 1, 1918. By 1882, the PRR became the largest railroad in the world (in terms of traffic and revenue) and the largest corporation in the world. During 1928–1938, the PRR electrified its New York–Washington, D.C. Line plus many of its commuter lines. During World War II, the PRR handled a large amount of freight and passengers. For most of its history, the Pennsylvania Railroad was a prosperous railroad.

New York Central Railroad

On August 13, 1831, the Mohawk & Hudson Railroad opened from Albany to Schenectady. Representatives from ten railroads in the Albany–Buffalo corridor met in Syracuse, New York, to form the New York Central Railroad on April 12, 1853. Railroad companies from Chicago, Illinois, to Erie, Pennsylvania, were consolidated into the Lake Shore & Michigan Southern Railway on April 6, 1869. The New York Central and Hudson River Railroad merged with the Lake Shore & Michigan Southern Railway on December 22, 1914 to form the New York Central Railroad. Completion of the St. Lawrence Seaway in 1959 resulted in a decline in railroad freight service. The building of the Interstate Highway system resulted in significant cuts in long distance and regional passenger service.

Penn Central Transportation Company

The Pennsylvania Railroad (PRR) and New York Central Railroad (NYC), rivals for much of the time, had much trackage not being used to capacity, and both had extensive passenger service. Both were vulnerable to competition from trucks, buses, and automobiles. The United States Congress passed the Federal-Aid Highway Act of 1956 authorizing construction of the Interstate Highway System, which was signed into law on June 29, 1956 by President Dwight David Eisenhower, who advocated the highway system construction so that U.S. troops could easily be transported in the event of a war. While the Interstate Highway System benefitted the trucking industry, and contributed to increased automobile usage, railroads paid taxes on their trackage and properties, and endured a sharp decline in railroad passenger ridership. On February 1, 1968, the Pennsylvania Railroad acquired the New York Central Railroad and became known as Penn Central Transportation Company, which was shortened to Penn Central Company on May 8, 1968. With differences in the management styles of the two railroads, the merger was not a success. Since the railroad's routes were not properly integrated, trains were lost on a regular basis. Industrial decline of the Northeast and Midwest resulted in a decline in freight carried. Decline in track maintenance reduced train speeds delaying shipments, increasing overtime costs, and operating costs. On June 21, 1970, Penn Central Company (PC) entered bankruptcy proceedings. In 1972, Hurricane Agnes destroyed many PC lines.

Erie Railroad

The New York & Erie Railroad began construction on their 6-foot broad-gauge line in New York State from the Hudson River at Piermont to Dunkirk on November 7, 1835, and the line was completed on May 9, 1851. The Atlantic & Great Western Railroad connected with the New York & Erie Railroad at Salamanca, New York. On June 25, 1861, the New York & Erie Railroad was reorganized as the Erie Railroad. On June 2, 1880, the work of converting the Erie Railroad to standard gauge was completed. In bankruptcy, the Atlantic & Great Western Railroad was reorganized into the New York, Pennsylvania & Ohio Railroad, later leased by the New York, Lake Erie & Western Railroad, and was bought by the Erie Railroad on November 3, 1895. The Erie Railroad held its own during the Depression until it entered bankruptcy on January 18, 1938. Its reorganization was successful and was soon able to pay dividends to its shareholders. The Erie prospered during the mid-1950s; however, it began an irreversible decline. The Erie's 1957 income was half of what it had in 1956, and by 1958 and 1959, the Erie had deficits. This led the Erie Railroad to do business with the Delaware, Lackawanna & Western Railroad (DL&W) which led to the elimination of duplicate freight facilities in Binghamton and Elmira. Between 1956 and 1957, the Erie shifted its passenger service from its Pavonia Terminal to the DL&W's newer Hoboken Terminal. The DL&W's main line between Binghamton and Elmira was taken out of service in favor of using the Erie's main line in 1958. The two railroads merged on October 17, 1960 into the Erie-Lackawanna Railroad.

Delaware, Lackawanna & Western Railroad

The Lackawanna & Western Railroad opened a line in Pennsylvania north from Scranton, Pennsylvania, to Great Bend on December 20, 1851. The Delaware & Cobb's Gap Railroad, chartered on December 4, 1850, built a line from Scranton east to the Delaware River. Before it opened, the Lackawanna & Western Railroad and Delaware & Cobb's Gap Railroad consolidated into the Delaware, Lackawanna & Western Railroad. The most profitable item shipped by the railroad was anthracite coal plus the Pocono Mountain Region served by this railroad became a popular vacation destination which generated large railroad passenger traffic. In August 1955, severe floods from Hurricane Diane destroyed 60 miles of track and closed the railroad for almost a month. The railroad merged with the Erie Railroad forming the Erie Lackawanna Railroad on October 17, 1960. Erie's large repair facility at Hornell, New York, closed and operations were transferred to Lackawanna's Scranton facility. The opening of interstate highways that paralleled much of the railroad diverted more traffic to trucks. The railroad petitioned to join Conrail and was legally conveyed into Conrail on April 1, 1976.

Erie Lackawanna Railway

After the Erie Railroad and Delaware, Lackawanna & Western Railway merged on October 17, 1960 to form the Erie Lackawanna Railroad, the railroad posted some profits through a reduction of parallel services and a steady reduction of long distance passenger service. The Erie Lackawanna Railway (EL) was formed on March 1, 1968, as a subsidiary of Dereco, a holding company of the Norfolk & Western Railway. In 1972, Hurricane Agnes destroyed much of the railroad's track with the cost of repairs and loss of revenues forcing the company into bankruptcy. The EL was accepted into Conrail.

Consolidated Rail Corporation (Conrail)

Under provisions of the Regional Rail Reorganization Act of 1973, Consolidated Rail Corporation (Conrail) was incorporated on October 25, 1974 and operations began on April 1, 1976 on PC, plus Erie Lackawanna, Lehigh Valley Railroad, Reading Company, Lehigh & Hudson River Railway, Central Railroad of New Jersey, and Pennsylvania–Reading Seashore Lines. Facing loss of market share to trucks, railroads and their unions asked the federal government for deregulation. The Staggers Rail Act of 1980 (named for Harley Staggers of West Virginia who chaired the House Committee on Energy & Commerce) signed into law by President Jimmy Carter on October 14, 1980 allowed a railroad to more easily abandon unprofitable rail lines and set its own freight rate. Under the Northeast Rail Service Act of 1981, Conrail was exempted from state taxes and required the secretary of transportation to make arrangements for the sale of the government's interest in Conrail. With legislation and the managerial improvements under L. Stanley Crane (Conrail chairman and CEO), Conrail became profitable. Norfolk Southern Railway acquired 58 percent (about 6,000 Conrail's route miles) and CSX acquired 42 percent (about 3,600 route miles). Operations under NS and CSX began on June 1, 1999.

Norfolk & Western Railway (N&W)

The earliest predecessor of the N&W was the City Point Railroad (CPRR), a 9-mile railroad formed in 1838 that extended from City Point to Hopewell, Virginia. In 1854, the CPRR became part of the South Side Railroad that connected Petersburg with Lynchburg. Construction of the Norfolk & Petersburg Railroad (N&P) began in 1853 and was completed in 1858. The N&P, South Side Railroad, and Virginia & Tennessee Railroad merged in 1870 to form the Atlantic, Mississippi & Ohio Railroad (AM&O) which extended 408 miles from Norfolk to Bristol, Virginia. In 1881, the AM&O was reorganized and renamed Norfolk & Western (N&W). The Flat-Top Coal Land Association (landowner of coal fields) was purchased by N&W and reorganized it as the Pocahontas Coal & Coke Company (PCCC). In 1886, N&W tracks were extended to the coal piers at Lambert's Point north of Norfolk, which became one of the busiest coal export facilities in the world. N&W's Roanoke Shops built their own steam locomotives. N&W's commitment to steam locomotives occurred because of its manufacturing capacity, skilled employees, and the fact that coal was the railroad's major commodity. N&W was the last major railroad to convert from steam to diesel power.

Virginian Railway

The Virginian Railway Company was the idea of civil engineer William Nelson Page and multi-millionaire Henry Huttleston Rogers. It began with the construction of the 4-mile Deepwater Railway in 1896 from an interchange with the Chesapeake & Ohio Railway to a sawmill at Robson, West Virginia. In 1898, it became the Deepwater Railway, which was extended to Matoaka, West Virginia. On March 8, 1907, the Deepwater Railway in West Virginia and Tidewater Railway in Virginia combined to form the Virginian Railway. The railroad was financed by Henry H. Rogers with no public debt. The railroad was completed on January 29, 1909, linking the coal mines of West Virginia to Hampton Roads, Virginia. The Virginian Railway operated more efficiently than its larger competitors, had new infrastructure, and no debt. A 134-mile portion of the railroad from Mullens, West Virginia, to Roanoke, Virginia, was electrified with an 11,000-volt catenary system. Pantograph equipped electric locomotives were supplied by Alco and Westinghouse. In June 1957, diesel locomotives replaced steam locomotives. The Virginian Railway was acquired by N&W on December 1, 1959.

Savannah & Atlanta Railway (S&A)

The Brinson Railway began construction in 1906 of a line from Savannah to Springfield which opened in 1907 and became the Savannah & Northwestern. On July 16, 1917, the Savannah & Atlanta (S&A) acquired the Savannah & Northwestern. On August 22, 1951, the Central of Georgia Railway purchased S&A, and the line became part of the Southern Railway.

Central of Georgia Railway

The Central Rail Road & Canal Company began construction in 1837 of a railroad from Savannah, which reached the eastern bank of the Ocmulgee River in 1843. It went into bankruptcy and reorganized as the Central of Georgia Railway (C of G) on November 1, 1865. During the Great Depression, the C of G went into receivership which lasted until 1948. In 1956, the St. Louis-San Francisco Railway (Frisco) gained control of the C of G; however, the Interstate Commerce Commission did not approve the merger, and the Frisco sold its C of G stock to the Southern Railway in 1963.

Southern Railway

The South Carolina Canal & Rail Road Company, chartered on December 19, 1827, operated the United States first regularly scheduled steam powered passenger train on December 25, 1830 via a 6-mile run from Charleston, South Carolina. In 1894, the Memphis & Charleston Railroad, Richmond & Danville, and the East Tennessee, Virginia & Georgia Railroad combined to form the Southern Railway. The Southern Railway combined with the Norfolk & Western Railway in 1980 and became the Norfolk Southern Railway with the Norfolk & Western Railway becoming a subsidiary to its system on June 1, 1982.

Georgia & Florida Railroad

Four short-line railroads formed the Georgia & Florida Railroad in 1906. The main line was from Madison, Florida, to Greenwood, South Carolina. It was acquired by the Southern Railway in February 1962 and was merged into the Southern's Central of Georgia (C of G) when the Southern Railway absorbed the C of G on June 1, 1971.

Wabash Railroad

Incorporated on August 31, 1852, the Lake Erie, Wabash, and St. Louis Railroad built a 242-mile railroad along the Wabash River primarily along the Wabash River in Indiana from the Ohio state line to the Illinois state line. The railroad merged on September 23, 1856 with the Toledo & Illinois to become the Toledo, Wabash, and Western Railroad (TW&W). Later mergers and reorganizations formed the Wabash, St. Louis & Pacific Railway on November 7, 1879 and the Wabash Railroad on August 1, 1889. The Wabash Railroad was sold at foreclosure on July 21, 1915 and reorganized as the Wabash Railway on October 22, 1915. Wabash was leased to the Norfolk & Western on October 16, 1964.

Pittsburgh & West Virginia Railway

Three railroads (Cross Creek Railroad; Pittsburgh, Toledo & Western Railroad; and Pittsburgh, Carnegie & Western Railroad) were consolidated into the Wabash Pittsburgh Terminal Railway (WPTR) on May 7, 1904; went into receivership on May 29, 1908; in August 1916, was sold at foreclosure; and reorganized in November 1916 as the Pittsburgh & West Virginia Railway (P&WV). On October 16, 1964, the Norfolk & Western acquired the New York Chicago & St. Louis Railroad (Nickel Plate) and leased a new W&LE on May 17, 1990, and the P&WV lease was transferred to the W&LE.

Akron, Canton & Youngstown Railway

The Akron & Youngstown Railway (AC&Y), founded in 1907, completed a 9.5-mile line from Mogadore to Akron, Ohio, in 1912. On March 1, 1920, the AC&Y leased the Northern Ohio Railway (NOR) which was later merged into the AC&Y. Norfolk & Western Railway (N&W) acquired control of the AC&Y on October 16, 1964 and operated it separately until it was merged into N&W on January 1, 1982.

New York Chicago & St. Louis Railroad

On April 13, 1881, the New York Chicago & St. Louis Railroad purchased the Buffalo, Cleveland & Chicago Railway. The Norwalk, Ohio, *Chronicle Newspaper* referred to this railroad as the "double track nickel-plated railroad," and it became known as the "Nickel Plate Road." Shortly after the railroad opened, William H. Vanderbilt, who owned the competing Lake Shore & Michigan Southern Railway (LS&MS), purchased the Nickel Plate on October 25, 1882 and allowed it to do enough business so that it would not fall into receivership. Since the Nickel Plate traversed Cleveland from east to west and had a natural route to the proposed Cleveland terminal, it was purchased by Oris Paxton Van Sweringen and his brother Mantis James Van Sweringen. LS&MS was controlled by Alfred Smith, a close friend of the Van Sweringens, who provided a vice president of the New York Central and some of his top managers to run the Nickel Plate Railroad for the Van Sweringens. The Nickel Plate merged into the Norfolk & Western on October 16, 1964.

Illinois Terminal Railroad

The Illinois Terminal Railroad was a successor to interurban railroads that were consolidated by William McKinley in the early 1900s into 550-mile Illinois Traction system that served central and southern Illinois. In 1937, the railroad reorganized into the Illinois Terminal Railroad (ITR). With more automobiles in use, ITR interurban passenger service became unprofitable. On March 3, 1956, ITR interurban passenger service ended followed by the last passenger line St. Louis–Granite City suburban line that ended service on June 22, 1958. ITR continued to operate diesel-powered freight service north and east of St. Louis up to 1981. Norfolk & Western (N&W) purchased its partners' interests in ITR on September 1, 1981. ITR officially merged into N&W on May 8, 1982. Active trackage in St. Louis was sold to Ironhorse Resources Inc. forming Railroad Switching Service of Missouri which operated freight service in St. Louis until June 21, 2004.

Wheeling & Lake Erie Railway

In 1880, the Wheeling & Lake Erie Railroad (W&LE) began standard-gauge operation on its mainline from Wheeling to Zanesville and Cleveland. It eventually connected Pittsburgh, Pennsylvania, with Toledo, Ohio. On December 1, 1949, the New York, Chicago, & St. Louis Railroad (Nickel Plate Road) leased the W&LE. When the Nickel Plate Road merged into the Norfolk & Western on October 16, 1964, W&LE was brought into N&W and later Norfolk Southern (NS). During 1990, NS sold most of the former W&LE to the Wheeling Acquisition Corporation, and the company was renamed Wheeling & Lake Erie Railroad on May 1, 1990.

Monongahela Railway

The Monongahela Railroad (a joint venture of the Pennsylvania Railroad and the Pittsburgh & Lake Erie Railroad) was consolidated with the Buckhannon & Northern Railway on July 1, 1915 to form the Monongahela Railway. On May 1, 1993, the Monongahela Railway was merged into the Consolidated Rail Corporation (Conrail). Monongahela's final locomotives (Eleven GE type B23-7R diesels) were divided between Norfolk Southern and CSX when they took over Conrail's operations.

Interstate Railroad

The Interstate Railroad was created in 1896 by the Virginia Coal & Iron Company to haul bituminous coal via a 6-mile rail line from Stonega to Appalachia where it connected with the Louisville & Nashville plus Virginia & Southwestern Railway. By 1929, it operated a 79.26-mile system in the state of Virginia. It was a well-engineered railroad with grades not exceeding 2.5 percent. It was taken over by the Southern Railway in October 1960, and disappeared into the Norfolk Southern Railway on October 31, 1985.

Reading Company

The Philadelphia & Reading Railroad (P&R) completed 93 miles of the double track connecting Pottsville via Reading to Philadelphia in 1843. In Philadelphia, the P&R built Port Richmond which allowed coal to be loaded onto ships for export, giving P&R almost full control of coal from mining to market. In 1900, shops were built in Reading, Pennsylvania, where the Reading was able to build its own locomotives. Larger steam locomotives such as the N1 class 2-8-8-2 were built to haul the increasing traffic. To simplify corporate structure, the Philadelphia & Reading Railroad became the Reading Company in 1924. The Reading operated a suburban commuter system out of Reading Terminal in Philadelphia that was later electrified. Dwindling coal shipping revenues, decline in freight revenues due to trucks carrying more freight over highways, and government regulations forcing the railroad to continue operating money-losing commuter service forced the Reading Company into bankruptcy in 1971. On April 1, 1976, the Reading Company sold its current railroad interest to Consolidated Rail Corporation. As of 1999, most former Reading lines are now part of Norfolk Southern Railway.

Atlantic, Mississippi & Ohio Railroad

The Norfolk & Petersburg Railroad was completed in 1858 in Virginia between Suffolk and Petersburg. The South Side Railroad in Virginia was completed in 1854 and connected City Point to Lynchburg. On October 1, 1856, the Virginia & Tennessee Railroad was completed from Lynchburg, Virginia, to Bristol, Tennessee. In 1870, these three railroads were combined into the Atlantic, Mississippi & Ohio Railroad (AM&O). In 1881, the AM&O was sold, and the new owners renamed it Norfolk & Western Railway which expanded and combined with the Southern Railway to form Norfolk Southern in 1982.

Central Railroad of New Jersey

The Central Railroad of New Jersey began in 1836 with the completion of the Elizabeth & Somerville Railroad in New Jersey in 1836. In 1852, the Somerville & Easton began service between Somerville and White House in 1848. On February 26, 1849, the companies merged and became the Central Railroad of New Jersey (CNJ). In 1893, America's first automatic, motor-operated semaphore signal was installed on the CNJ near Phillipsburg, New Jersey. The Reading Railroad gained control of the CNJ in 1901 which lasted until the creation of Conrail on April 1, 1976. In 1925, America's first successful commercial diesel locomotive CNJ No. 1000 was placed in service at Bronx Terminal. CNJ entered a ten-year bankruptcy in 1939. CNJ's traffic base declined as manufacturing moved out of the region plus more trucks on expanding highways resulted in a decline in railroad freight business. Cancellation of mail contracts in 1965 resulted in the elimination of eighteen CNJ passenger trains. CNJ was absorbed into Conrail on April 1, 1976.

Lehigh Valley Railroad

Construction of the Lehigh Valley Railroad (LVR) began in 1853, and the line opened between Easton and Allentown, Pennsylvania, on June 11, 1855 followed by the section between Allentown and Mauch Chunk on September 12, 1855. Asa Packer became president of the LVR on January 13, 1862. By 1866, the LVR reached Wilkes Barre, Pennsylvania. In 1867, the LVR was completed from Wilkes Barre, Pennsylvania, to Waverly, New York, where coal was transferred to the 6-foot broad-gauge Erie Railroad. A third rail was installed on Erie mainline tracks to allow LVR cars to run through to Elmira and later to Buffalo, New York. By 1893, the LVR acquired or controlled 53,000 acres of coal land, making it possible to mine as well as transport coal. Following the Great Depression, the LVR went into a decline. Passenger ridership declined as more people purchased automobiles. In the 1950s, the opening of interstate highways helped the trucking industry offer door to door service, and completion of the St. Lawrence Seaway allowed grain shipments to bypass railroads and go directly to overseas markets. In 1962, the Pennsylvania Railroad requested and received ICC approval to acquire control of the LVR. The February 1, 1968 merger of the Pennsylvania Railroad and New York Central Railroad into Penn Central failed on June 21, 1970. The LVR declared bankruptcy on July 24, 1970. On April 1, 1976, the LVR was merged into the Consolidated Rail Corporation (Conrail). The former LVR Lehigh Line from Port Reading Junction in Manville, New Jersey, to Penn Haven Junction in Lehigh Township, Pennsylvania, is operated by Norfolk Southern Railway (NS). CSX operates from Port Reading Junction in Manville, New Jersey, to Packerton and Lehighton, Pennsylvania. There is also Conrail Shared Assets Operation in North Jersey and South Jersey/Philadelphia that serves as a local carrier and switching company for its owners CSX Transportation and Norfolk Southern Railway. When Conrail was split up between CSX and NS, certain track was kept separate to avoid giving one railroad an advantage over the other railroad.

Pennsylvania Railroad K4 "Pacific" steam locomotive No. 1361 with a 4-6-2 wheel arrangement (built by the Pennsylvania Railroad's Juniata Shops in May 1918) is at South Amboy, New Jersey, on August 4, 1954. This locomotive handled passenger and mail trains in Pennsylvania, plus commuter trains in New Jersey, until retired from revenue service in 1956. It was cosmetically refurbished and placed on display at Horseshoe Curve on June 8, 1957. It was moved to the Railroaders Memorial Museum in Altoona on September 5, 1985. Type GP9 diesel No. 7048 replaced the steam locomotive at Horseshoe Curve on September 16, 1985. (*Photographer Jim Fritts Used with permission from Audio-Visual Designs [www. audiovisualdesigns.com]*)

In 1926, Pennsylvania Railroad Mountain type locomotive No. 6903 (with a 4-8-2 wheel arrangement built by Baldwin Locomotive Works in 1926) is powering a train of reefer cars through the Borough of Mount Union in Huntingdon County, Pennsylvania. This locomotive featured a large cross air tank at the front of the locomotive above the pilot. (*Photographer Stephen D. Bogen. Used with permission from Audio-Visual Designs [www. audiovisualdesigns.com]*)

Pennsylvania Railroad diesel locomotive No. 5827 (Baldwin Locomotive Works type DR-12-8-1500/2 informally known as a Centipede built in October 1947) is at East Altoona, Pennsylvania on October 28, 1959. DR stands for diesel road locomotive, 12 stands for 12 axles of which 8 were driven, and 2 engines of 1,500 horsepower each. The nickname Centipede came from the numerous axles almost like the legs of a centipede. (*Photographer William Volkmer. Used with permission from Audio-Visual Designs [www.audiovisualdesigns.com]*)

In August 1964, Pennsylvania Railroad 1,500-horsepower diesel electric locomotive No. 9815 (EMD type F7 built in August 1964 and equipped with a radio antenna on the roof) is at Jefferson, Indiana. (*Photographer Mac Owen collection. Used with permission from Audio-Visual Designs [www.audiovisualdesigns.com]*)

Model 660 "Doodlebug" (name derived from the slow speed at which this type of vehicle would doodle down the track) No. 4653 built by J.G. Brill in 1928, chartered by the Lehigh Valley Chapter of the National Railway Historical Society, is at Lambertville, New Jersey, on November 3, 1957. This type of vehicle was used by some railroads to provide passenger and mail service on lightly used branch lines. (*Photographer Douglas C. Wornom. Used with permission from Audio-Visual Designs [www.audiovisualdesigns.com]*)

On May 18, 1977, Amtrak GG1 No. 4935 (built by the Pennsylvania Railroad's Juniata shops at Altoona in March 1943) is making a station stop at Trenton, New Jersey. As the Pennsylvania Railroad was involved with an extensive electrification project in the 1930s, the GG1 was chosen as its major electric locomotive. This locomotive was restored in 1977 to its 1943 appearance with funding provided by National Railway Historical Society chapters. The locomotive has been preserved at the Railroad Museum of Pennsylvania at Strasburg, Pennsylvania. (*Photographer William Rosenberg. Used with permission from Audio-Visual Designs [www.audiovisualdesigns.com]*)

Magnificent Bicentennial colors are on Conrail GG1 No. 4800 (built in August 1934) at Harrisburg, Pennsylvania, on April 12, 1976. Between 1934 and 1943, there were 139 GG1 locomotives built, of which fifteen were built by General Electric (GE Erie, PA) and 124 by the Pennsylvania Railroad (PRR Altoona, PA). On January 28, 1935, PRR ran a special train powered by GG1 No. 4800 marking the completion of electrification of the Washington, D.C., to New York City line. (*Photographer William Rosenberg. Used with permission from Audio-Visual Designs [www.audiovisualdesigns.com]*)

On May 19, 1969, Penn Central GG1 No. 4902 (built by the Pennsylvania Railroad's Juniata shops at Altoona in May 1940) has coupled on to the Golden Spike Centennial Limited at Baltimore, Maryland, for the last lap of the trip to Pennsylvania Station in New York City. This commemorated the 100th anniversary of the driving of the Golden Spike at Promontory Summit, Utah, on May 10, 1869 where the rails of the Central Pacific Railroad and Union Pacific Railroad were joined. The special train was organized by Ross Rowland, Jr., and his High Iron Company. (*Photographer Don Wood. Used with permission from Audio-Visual Designs [www.audiovisualdesigns.com]*)

Hudson (named after the Hudson River) type 4-6-4 steam locomotive No. 5214 (built in 1927 by American Locomotive Company [Alco]) is at Harmon, New York, on June 8, 1952. Between 1927 and 1931, there were 205 of these class J-1 locomotives (Nos. 5200–5404) built by Alco for the New York Central Railroad. Built for high-speed passenger service, these locomotives powered New York Central's major passenger trains such as the *20th Century Limited* and the *Empire State Express*. (*Photographer Bob Collins. Used with permission from Audio-Visual Designs [www.audiovisualdesigns.com]*)

Six New York Central Railroad diesel locomotives headed by No. 3821 (1,625-horsepower type RF-16A with a "Sharknose" body style built by Baldwin-Lima-Hamilton Corporation in February 1952) are powering a freight train past the tower at Bellefontaine, Ohio. (*Photographer Jim Hawk. Used with permission from Audio-Visual Designs [www.audiovisualdesigns.com]*)

On August 21, 1961, in Chicago, an Indiana Harbor diesel switcher is passing New York Central Railroad 1,600-horsepower diesel locomotive No. 1053 (type FA2 built by Alco in June 1951). (*Photographer R. R. Wallin. Used with permission from Audio-Visual Designs [www.audiovisualdesigns.com]*)

Reading Railroad four-axle 2,500-horsepower diesel locomotive No. 3623 (EMD type GP35 built in December 1963) is traversing a route through the rugged mountains of Pennsylvania. (*Used with permission from Audio-Visual Designs [www.audiovisualdesigns.com]*)

On Thanksgiving Day November 23, 1972, Lehigh Valley Railroad four-axle 2,000-horsepower diesel locomotive No. 319 (EMD type GP38-2 built in November 1972) is waiting for the next assignment at the railroad's Newark, New Jersey, railroad yard. The Lehigh Valley Railroad had twelve of these locomotives (Nos. 314–325). (*Photographer Max Robin. Used with permission from Audio-Visual Designs [www.audiovisualdesigns.com]*)

During the winter of 1959, Lehigh Valley Railroad 2,000-horsepower diesel locomotive No. 602 (Alco type PA1 built in March 1948) is westbound making a station stop at Lehighton, Pennsylvania, leading the famed *Black Diamond* passenger train. This train (named for the railroad's major cargo, anthracite coal) operated from New York City to Buffalo from 1896 to May 11, 1959. (*Photographer Douglas Wornom. Used with permission from Audio-Visual Designs [www.audiovisualdesigns.com]*)

Monongahela Railway diesel locomotive No. 1205 (originally built for the New York Central Railroad as No. 3805 [Baldwin locomotive Works type RF-16] in December 1951) is at Brownsville, Pennsylvania, on September 21, 1971. *(Photographer Dan Cupper. Used with permission from Audio-Visual Designs [www.audiovisualdesigns.com])*

In October 1956, Jersey Central Lines (also known as Central Railroad of New Jersey) 2,400-horsepower six-axle diesel locomotive No. 2408 (Fairbanks Morse type H-24-66 also known as Trainmaster was built in March 1956) is at Bay Head, New Jersey, on the New York & Long Branch Railroad. This locomotive was noted for its pulling power and rapid acceleration. *(Photographer Don Wood. Used with permission from Audio-Visual Designs [www.audiovisualdesigns.com])*

Delaware Lackawanna & Western Railroad (DL&W) 1,350-horsepower type FTA diesel locomotive No. 602 built by EMD in April 1945 is one of twelve A units purchased by that railroad. It is believed that this successful model diesel locomotive initiated the eventual end of the steam locomotive in the United States. (*Used with permission from Audio-Visual Designs [www.audiovisualdesigns.com]*)

EMD 600-horsepower diesel electric switcher No. 430 type SW1 built in April 1940 (S for 600 horsepower, W for welded, and the number 1 was added to distinguish the new design from the earlier EMD SW) is one of eleven units (Nos. 427–437) purchased by the Delaware Lackawanna & Western Railroad. (*Used with permission from Audio-Visual Designs [www.audiovisualdesigns.com]*)

In 1964, Wabash Railroad six-axle 2,400-horsepower diesel electric locomotive No. 597 (Fairbanks Morse and Canadian Locomotive Company type H-24-66 built in 1956) is posing for its official portrait at the Alco yard in Schenectady, New York, after it was re-powered with a model 252 engine in 1964. *(Used with permission from Audio-Visual Designs [www.audiovisualdesigns.com])*

On April 27, 1975, Southern Railway 2-8-2 "Mikado" steam locomotive No. 4501 steam locomotive (built in October 1911 by Baldwin Locomotive Works) is ready to depart the Knoxville, Tennessee, station for a National Railway Historical Society excursion. In 1948, this locomotive was retired from the Southern Railway, sold to the Kentucky & Tennessee Railway in 1948, purchased by Paul H. Merriman in 1964, returned to the Southern Railway steam program in 1966, and operated until 1994 when the Norfolk Southern discontinued the steam program due to cost increases. In later years, the locomotive has been used on Tennessee Valley Railroad Museum excursions. *(Photographer Joe Kmetz. Used with permission from Audio-Visual Designs [www.audiovisualdesigns.com])*

Southern Railway 1,350-horsepower diesel electric locomotive No. 6100 is on display at EMD's fiftieth anniversary celebration on September 9, 1972 at McCook, Illinois. This locomotive was EMD's model FT demonstrator No. 1030, built in May 1939, and ran with a companion B unit. The Southern Railway had thirty-six FT A units (Nos. 4100–4127, 6101–6104, and 6800–6803). *(Used with permission from Audio-Visual Designs [www.audiovisualdesigns.com])*

In June 1971, Erie Lackawanna Railway six-axle 1,500-horsepower diesel electric locomotive No. 1153 (type DRS-6-6-1500 built by the Baldwin Locomotive Works in September 1950 for the Erie Railroad) is shown at Hornell, New York. (*Photographer Jim Shaughnessy. Used with permission from Audio-Visual Designs [www.audiovisualdesigns.com]*)

Wabash Railroad steam locomotive No. 2911 is at Brooklyn, Illinois, in July 1949. These twenty-five "Northern" 4-8-4 locomotives class 0-1 (Nos. 2900–2924) were built by Baldwin Locomotive Works with eighteen delivered in 1930 and the remainder received in 1931. By February 1956, all of these locomotives had been retired. This locomotive weighed 454,090 pounds, and the loaded weight of the tender was 281,000 pounds. Tractive effort (force the locomotive can generate for hauling the load) was 70,817 pounds. *(R. R.Wallin collection. Used with permission from Audio-Visual Designs [www.audiovisualdesigns.com])*

On October 12, 1960, Wabash Railroad 2,250-horsepower diesel electric locomotive No. 1007 (EMD type E8 built in November 1949) is at Decatur, Illinois. *(R. R.Wallin collection. Used with permission from Audio-Visual Designs [www.audiovisualdesigns.com])*

Decatur, Illinois, is the location of Wabash Railroad diesel locomotive No. 807 on July 28, 1961. This was one of ten (Nos. 805–814) Alco type FA locomotives (F denoting these locomotives were geared primarily for freight service) built during January to March 1949. *(Photographer R. R.Wallin. Used with permission from Audio-Visual Designs [www.audiovisualdesigns.com])*

Virginian 2-8-4 "Berkshire" steam locomotive No. 507 built by Lima Locomotive Works is on display at Roanoke, Virginia, in 1957. This was one of five class BA locomotives (Nos. 505–509) built by Lima in 1946. The locomotive weighed 460,400 pounds, and the loaded tender weighed 410,200 pounds. *(Photographer Richard J. Cook. Used with permission from Audio-Visual Designs [www.audiovisualdesigns.com])*

In 1957, Virginian Railway class AG 2-6-6-6 steam locomotive No. 903 (one of eight [Nos. 900–907] known as Blue Ridge type of locomotives delivered between March and June 1945 by Lima Locomotive Works) is at Roanoke, Virginia. These locomotives handled coal trains well into the 1950s; they were retired by 1955 and scrapped by 1960. The Virginian and Chesapeake & Ohio were the only two United States railroads that bought the 2-6-6-6. (*Photographer Richard J. Cook. Used with permission from Audio-Visual Designs [www.audiovisualdesigns.com]*)

Virginian Railway 3,300-horsepower class EL-C rectifier unit No. 132 (one of twelve electric locomotives [Nos. 130–141] delivered between October 1956 and January 1957 by General Electric Company) is powering a National Railway Historical Society special train using Norfolk & Western Railway coaches east into the yard at Roanoke, Virginia, in 1957. (*Photographer Richard J. Cook. Used with permission from Audio-Visual Designs [www.audiovisualdesigns.com]*)

In September 1958, New York Chicago & St. Louis Railroad, commonly known as the Nickel Plate, Berkshire steam locomotive class S-2 No. 763 (built in September 1944 and one of thirty class 2 [Nos. 740–769] built by Lima Locomotive Works) is fresh out of the Conneaut, Ohio, shops awaiting a break in run. This locomotive has been preserved at the Age of Steam Roundhouse Museum at Sugarcreek, Ohio. (*Photographer Don Wood. Used with permission from Audio-Visual Designs [www.audiovisualdesigns.com]*)

New York Chicago & St. Louis Railroad, commonly known as the Nickel Plate, Berkshire steam locomotive class S-4 No. 817 (originally Wheeling & Lake Erie Railway class K-1 No. 6417 built by Alco in 1942) is taking on water at the village Dillonvale in Jefferson County, Ohio, on June 16, 1955. (*Photographer Bob Collins. Used with permission from Audio-Visual Designs [www.audiovisualdesigns.com]*)

This is a broadside view of stylish blue and white "Bluebird" paint scheme Nickel Plate 2,000-horsepower passenger diesel locomotive No. 185 (type PA1 built in December 1947 by Alco in Schenectady, New York, in a partnership with GE). (*Used with permission from Audio-Visual Designs [www.audiovisualdesigns.com]*)

Norfolk & Western Railway class A 2-6-6-4 steam locomotive No. 1218 (one of forty-three [Nos. 1200–1242] built by N&W in 1936 with an engine weight of 573,000 pounds, tender loaded weight of 378,600 pounds, and a tractive effort of 125,897 pounds) and class J steam locomotive No. 611 (one of fourteen [Nos. 600–613] built by N&W in 1943 with an engine weight of 494,000 pounds, loaded tender weight of 378,600 pounds, and a tractive effort of 77,899 pounds) are climbing the grade between Roanoke and Christiansburg, Virginia, during a double run-by on the National Railway Historical Society August 1, 1987 Convention excursion. (*Photographer Carl H. Sturner. Used with permission from Audio-Visual Designs [www.audiovisualdesigns.com]*)

On December 8, 1949, the Norfolk & Western Railroad streamlined passenger train Powhatan Arrow is near Singer, Virginia. This train began operation on April 28, 1946 over the 676-mile route connecting Norfolk, Virginia, with Cincinnati, Ohio. In 1949, the train was re-equipped with streamlined cars from Pullman-Standard painted Tuscan red with gold trim and a dark brown roof. It made its final run in the spring of 1969. (*Used with permission from Audio-Visual Designs [www.audiovisualdesigns.com]*)

Norfolk & Western Railway four-axle 2,000-horsepower diesel electric locomotive No. 4128 (EMD type GP38AC [basically a GP38 with an AR10 alternator instead of a GP38 GM built D32 generator for increased tractive effort] built in June 1971) is shown after delivery at Bellevue, Ohio, on June 27, 1971. (*Photographer Howard W. Ameling. Used with permission from Audio-Visual Designs [www.audiovisualdesigns.com]*)

On May 1, 1974, Norfolk & Western Railway (N&W) six-axle 3,300-horsepower diesel electric locomotive No. 1776 (EMD type SD45 built in February 1970) is on its first revenue trip after being painted in a special Bicentennial color scheme followed by four-axle 1,750-horsepower GP9 locomotives No. 710, built in September 1955, and No. 914 built in September 1959. *(Photographer David L. Foster. Used with permission from Audio-Visual Designs [www.audiovisualdesigns.com])*

In September 1954, the fireman looks to the rear as the engineer eases J-1 class N&W 4-8-4 "Northern" steam locomotive No. 610 (one of six [Nos. 605–610] built in 1944 by N&W) down to the Roanoke, Virginia, station to couple up to its train. There were fourteen of these locomotives (Nos. 600–613) that had an engine weight of 378,600 pounds, loaded tender weight of 378,600 pounds, and a tractive effort of 77,899 pounds. *(Photographer Marvin H. Cohen. Used with permission from Audio-Visual Designs [www.audiovisualdesigns.com])*

N&W class A articulated 2-6-6-4 steam locomotive No. 1218 (one of forty-three locomotives [Nos. 1200–1242] built by N&W in 1936) is on a rail excursion on 19th Street in Erie, Pennsylvania, on August 18, 1990. In July 1959, No. 1218 was retired. In 2001, the Norfolk Southern donated No. 1218 to the City of Roanoke. On April 2, 2012, the City of Roanoke donated both Nos. 1218 and 611 to the Virginia Museum of Transportation. (*Kenneth C. Springirth photograph*)

On June 10, 1989, N&W No. 611 (one of three J-1 class N&W 4-8-4 "Northern" steam locomotives [Nos. 611–613] built in 1950 by N&W) is at North East, Pennsylvania, for an eastbound rail excursion to Buffalo, New York. The J-class locomotives were noted for their dual service capability because of their ability to handle long freight trains and could run a passenger train at 90 miles per hour. (*Kenneth C. Springirth photograph*)

Spectators are watching Norfolk & Western steam locomotive No. 611 posing for a rail excursion photo stop on 19th Street at Myrtle Street in Erie, Pennsylvania, on August 8, 1992. This was one of fourteen locomotives that featured the maximum use of roller bearings on axles and cadmium plated main side rods. (*Kenneth C. Springirth photograph*)

On August 18, 1990, Norfolk Southern Railway (NS) six-axle 3,900-horsepower diesel electric locomotive No. 8612 (GE type C39-8 built in August 1985 one of 114 [Nos. 8550–8663] all of which have been retired) is leading a freight train on 19th Street at Sassafras Street in Erie, Pennsylvania. On September 27, 2001, the last NS freight train operated through Erie via the 19th Street tracks. On Friday October 12, 2001, a special train made the last run for invited guests through the heart of Erie on the 19th Street tracks. NS rerouted 5 miles of its rail line along the CSX railroad right of way between East and Pittsburgh Avenues. (*Kenneth C. Springirth photograph*)

2

Multicolored Norfolk Southern Railway

As noted by the Norfolk Southern Railway July 4, 2012 news release, "Norfolk Southern Railway has honored its predecessor railroads during 2012, its 30th anniversary year, by painting 20 new locomotives in commemorative schemes that reflect the heritage of those predecessors." Over the years, hundreds of railroad companies were merged, reorganized, and consolidated into the Norfolk Southern Railway that itself was created from the consolidation of the Southern Railway and Norfolk & Western Railway on June 1, 1982 and expanded its heritage by acquiring a portion of Conrail and beginning operation on it on June 1, 1999. Wick Moorman, chief executive officer of the Norfolk Southern Railway (NS), noted in a March 1, 2012 news release, "The heritage locomotives reflect the pride we take in our long and colorful history. As they travel through our system, these state of the art units in vintage livery will serve as reminders to our customers, employees, and communities that the modern rail network that keeps America competitive today and into the future has deep roots in the nation's past." Staying as true as possible to the original paint schemes, each print scheme was designed to fit the contemporary locomotives while endeavoring to reflect the original designs. While the NS employees at Altoona, Pennsylvania, and Chattanooga, Tennessee, painted GE ES44AC locomotives, the EMD SD70ACe locomotives were painted at the Progress Rail Services Facility in Muncie, Indiana. Upon completion, these locomotives were used to haul freight across the rail system.

NS six-axle 4,300-horsepower locomotive No. 1065 (EMD type SD70ACe painted and placed in service in the Savannah& Atlanta Railway heritage paint scheme at Muncie, Indiana, on May 31, 2012) is at Dyers Quarry (supplier of crushed aggregates served by NS) in the borough of Birdsboro in Berks County, Pennsylvania, in May 2020. (*Brian A. Keates photograph*)

On September 8, 2020, NS locomotive No. 1066 (EMD type SD70ACe painted in the New York Central Railroad [NYC] heritage paint scheme at Muncie, Indiana, and placed in service on May 31, 2012) is trailing on an eastbound intermodal train thru the borough of Wyomissing in Berks County, Pennsylvania. This is the NS Harrisburg Line that was formerly two Reading Company Lines: the main line from Philadelphia to Reading and the Lebanon Valley Branch between Reading and Harrisburg. (*Brian A. Keates photograph*)

Trailing on an eastbound coal train, on April 5, 2021, NS locomotive No. 1067 (EMD type SD70ACe painted in the Reading Company heritage paint scheme at Muncie, Indiana, and placed in service June 19, 2012) is at the Brickyard Crossing west of the city of Altoona in Blair County, Pennsylvania. (*Brian A. Keates photograph*)

NS locomotive No. 1068 (EMD type SD70ACe painted in its Erie Railroad heritage paint scheme at Muncie, Indiana, and placed in service on June 19, 2012) is leading an eastbound general manifest train (typically having a variety of cars such as boxcars, flatcars, gondolas, and tank cars carrying a variety of commodities) through Altoona, Pennsylvania, on May 4, 2020. (*Brian A. Keates photograph*)

On May 22, 2022, NS locomotive No. 1069 (EMD type SD70ACE painted in the Virginian Railway heritage paint scheme at Muncie, Indiana, and placed in service on June 27, 2012) is one of the helper locomotives on an eastbound intermodal train at Gallitzin, Pennsylvania. (*Brian A. Keates photograph*)

Wyomissing, Pennsylvania, is the location on October 4, 2020 of NS locomotive No. 1070 (EMD type SD70ACe painted in the Wabash heritage paint scheme at Muncie, Indiana, and placed in service on June 27, 2012). In response to the song "Wabash Cannonball," the Wabash Railroad in 1950 renamed its daytime train between Detroit, Michigan and St. Louis, Missouri, the *Wabash Cannon Ball*, which was discontinued when Amtrak took passenger service in May 1971. (*Brian A. Keates photograph*)

On January 1, 2019, NS locomotive No. 1071 (EMD type SD70ACe painted in the Jersey Central Lines [CNJ] heritage paint scheme at Muncie, Indiana, and placed in service on June 27, 2012) is trailing on a westbound general manifest train at Cove in Perry County, Pennsylvania. In 1925, CNJ purchased the first commercially successful diesel electric locomotive No. 1000 that was built by GE in partnership with Ingersoll-Rand and Alco, which served for over thirty years in the Bronx, New York, yards. (*Brian A. Keates photograph*)

In this picturesque spot at Cove, Pennsylvania, on May 31, 2021, NS locomotive No. 1072 (EMD type SD70ACe painted in the Illinois Terminal Railroad [ITR] paint scheme at Muncie, Indiana, and placed in service on June 27, 2012) is leading a westbound intermodal train. The ITR 1940 System Map stated "550 miles of Steam-Electric Service." This was second to Pacific Electric Railway Company which was the largest in the world with over 1,000 miles of track. (*Brian A. Keates photograph*)

On April 30, 2022, NS locomotive No. 1073 (EMD type SD70ACe painted in the Penn Central heritage paint scheme at Muncie, Indiana, and placed in service on June 27, 2012) is leading a westbound intermodal train passing the Altoona, Pennsylvania, transportation center. (*Brian A. Keates photograph*)

Trailing on a westbound intermodal train, NS locomotive No. 1074 (EMD type SD70ACe painted in the Delaware, Lackawanna & Western heritage paint scheme at Muncie, Indiana, and placed in service on June 27, 2012) is passing through Cove, Pennsylvania, on December 28, 2017. (*Brian A. Keates photograph*)

On May 11, 2022, mid-train distributed power unit (DPU) 4,400-horsepower NS locomotive No. 8025 (GE type ES44AC painted in the Monongahela Railway heritage paint scheme at Chattanooga, Tennessee, on June 15, 2012) is on an eastbound intermodal train near the Allegheny Tunnel at Gallitzin, Pennsylvania. A DPU (remotely controlled by the operator in the leading locomotive increases both pulling and stopping of a longer train) is one or two extra locomotives placed at an intermediate point such as in the middle of the train or it can be at the end of the train. (*Brian A. Keates photograph*)

On June 30, 2021, Wyomissing, Pennsylvania, is the location of a westbound intermodal train with trailing NS locomotive No. 8098 (GE type ES44AC painted in the Consolidated Rail Corporation [Conrail] heritage paint scheme at Juniata, Pennsylvania, on March 15, 2012). Conrail continues to do business as a service provider in Shared Assets Areas that were excluded from the division of its operations during its acquisition by Norfolk Southern Railway and CSX Corporation. (*Brian A. Keates photograph*)

A westbound intermodal train is passing through Wyomissing, Pennsylvania on January 17, 2021 with trailing NS locomotive No. 8099 (GE type ES44AC painted in the Southern Railway heritage paint scheme at Chattanooga, Tennessee, on March 20, 2012). The Southern originated as the South Carolina Canal & Rail Road Company, which operated the first regularly scheduled steam passenger train in the United States on Christmas Day, December 25, 1830 on a 6-mile line out of Charleston, South Carolina. (*Brian A. Keates photograph*)

On May 23, 2020, trailing on a westbound intermodal train that has emerged from the Allegheny Tunnel at Gallitzin, Pennsylvania, is NS locomotive No. 8100 (GE type ES44AC painted in the New York Chicago & St. Louis Railroad [Nickel Plate Road] heritage paint scheme at Juniata, Pennsylvania, on March 29, 2012). The Nickel Plate was famous for its 2-8-4 Berkshire steam locomotives, with fifteen (Nos. 700–714) purchased from American Locomotive Company and sixty-five (Nos. 715–779) purchased from Lima Locomotive Works. This locomotive type was noted as "The greatest 2-8-4 that was ever built." (*Brian A. Keates photograph*)

NS locomotive No. 8101 (GE type ES44AC painted in the Central of Georgia Railway heritage paint scheme at Chattanooga, Tennessee, on April 10, 2012) is trailing on an intermodal train through Cove, Pennsylvania, on May 21, 2020. The Central of Georgia Railway from July 17, 1947 to April 30, 1971 operated the *Nancy Hanks* passenger train in Georgia connecting Atlanta with Savannah. It was named for a racehorse that was named for Abraham Lincoln's mother Nancy Hanks Lincoln. (*Brian A. Keates photograph*)

Awaiting the next assignment is NS locomotive No. 8102 (GE type ES44AC painted in the Pennsylvania Railroad heritage paint scheme at Juniata, Pennsylvania, on April 18, 2012) at West Fitchburg, Massachusetts, in 2016. The Pennsylvania Railroad was named for the state of Pennsylvania where it was established. By 1882, the Pennsylvania Railroad was the largest railroad (by traffic and revenue) and the largest corporation in the world. (*Robert A. Elder photograph*)

On a cold January 2021, NS locomotive No. 8103 (GE type ES44AC painted in the Norfolk & Western Railway [N&W] heritage paint scheme at Juniata, Pennsylvania, on April 9, 2012) is trailing on a westbound intermodal train at Wyomissing, Pennsylvania. The N&W built its own steam locomotives at their Roanoke, Virginia, shops. The class A, J and Y6 locomotives brought the company fame for their excellence in steam locomotives. (*Brian A. Keates photograph*)

Trailing on an eastbound NS intermodal train through Wyomissing, Pennsylvania, on October 17, 2020 is NS locomotive No. 8104 (GE type ES44AC painted in the Lehigh Valley Railroad [LV] heritage paint scheme at Juniata, Pennsylvania, on April 23, 2012). The flagship passenger train of the LV was the *Black Diamond* (named after the anthracite coal the LV transported) that connected New York City to Buffalo from May 18, 1896 to May 11, 1959. (*Brian A. Keates photograph*)

September 16, 2018, trailing on an NS intermodal train through Cove, Pennsylvania, is NS locomotive No. 8105 (GE type ES44AC painted in the Interstate Railroad cream, orange, and gray heritage paint scheme at Chattanooga, Tennessee, on April 19, 2012). In spite of its name, this railroad operated entirely within the state of Virginia. (*Brian A. Keates photograph*)

February 2, 2020, trailing on a westbound NS intermodal train through Cove, Pennsylvania, is NS locomotive No. 8114 (GE type ES44AC painted in the original NS heritage paint scheme at Chattanooga, Tennessee, on April 25, 2012). Not to be confused with today's Norfolk Southern, this was a railroad that operated from Norfolk, Virginia, to Charlotte, North Carolina. It was acquired by the Southern Railway on January 1, 1974. (*Brian A. Keates photograph*)

Stored at the NS Juniata Shops in Altoona on October 24, 2021, NS 4,300-horsepower locomotive No. 1801 EMD type SD70ACC is awaiting its next assignment. This locomotive was originally No. 2548, built in September 1994 as 4,000-horsepower EMD type SD70. It was rebuilt to type SD70ACC, renumbered 1801, and painted in a special yellow, gray, black, and white paint scheme to highlight the conversion from D.C. to AC by Progress Rail at Muncie, Indiana, in March 2018. (*Brian A. Keates photograph*)

On October 19, 2020, NS 4,400-horsepower diesel locomotive No. 4000 type AC44C6M is leading a westbound intermodal train through Annville Township in Lebanon County, Pennsylvania. This locomotive was originally built by GE as 4,000-horsepower No. 8799 Dash 9-40C in January 1995. It was rebuilt from DC to AC traction by American Motive Power Inc., renumbered 4000, and painted blue, gray, gold, white, black with DC to AC lettering and AC wave form under the cab numbers on August 1, 2015. (*Brian A. Keates photograph*)

On February 22, 2021, single NS locomotive No. 4001 type AC44C6M is powering an eastbound intermodal train through Cove, Pennsylvania. This locomotive was originally GE type Dash 9-40C No. 8879 built in March 1995. American Motive Power Inc. rebuilt this locomotive to type AC44C6M on August 1, 2015 with the same paint scheme as on No. 4000. (*Brian A. Keates photograph*)

On March 31, 2018, NS locomotive No. 4002 is trailing on a westbound intermodal train at Cove, Pennsylvania. Originally this locomotive was GE type Dash 9-40C No. 8789 built in January 1995. NS Roanoke Locomotive Shop rebuilt this locomotive to type AC44C6M on September 11, 2016 painted in special black, Tuscan red, gray, and white paint scheme with DC to AC lettering and AC wave form under the cab numbers. *(Brian A. Keates photograph)*

An eastbound intermodal train is at the Brickyard crossing in Altoona, Pennsylvania, on June 7, 2021 with NS locomotive No. 4003. This locomotive originally was GE type Dash 9-40C No. 8839 built in February 1995. NS Roanoke locomotive shop rebuilt this locomotive to type AC44C6M on October 19, 2016 painted in special black, Tuscan red, gray, and white paint scheme with DC to AC lettering and AC wave form under the cab numbers. *(Brian A. Keates photograph)*

On February 27, 2021, NS locomotive No. 4004 is powering a westbound intermodal train past the Altoona (Pennsylvania) Transportation Center that opened on October 23, 1986 for local bus, intercity bus, and rail passenger service. Originally this locomotive was GE type Dash 9-40C No. 8866 built in March 1995. NS Juniata Locomotive Shop rebuilt this locomotive to type AC44C6M on September 16, 2016 painted in special black, blue, gray, and white paint scheme with DC to AC lettering and AC waveform under the cab numbers. (*Brian A. Keates photograph*)

NS locomotive No. 4005 is trailing on a westbound intermodal train through Wyomissing, Pennsylvania, on July 26, 2022. Originally this locomotive was GE type Dash 9-40C No. 8867 built in March 1995. NS Juniata shop rebuilt this locomotive to type AC44C6M on December 12, 2016 painted in special black, blue, gray, and white paint scheme with DC to AC lettering and AC wave form under the cab numbers. (*Brian A. Keates photograph*)

On April 21, 2021, 3,000-horsepower NS locomotive No. 4655 is switching cars for local train H31 at Sinking Spring, Pennsylvania. This locomotive originally was NS No. 4637 EMD type GP59 built in December 1989. It was rebuilt by NS on January 30, 2014 into type GP59E painted in special Operation Lifesaver paint scheme. (*Brian A. Keates photograph*)

NS 2,000-horsepower locomotive No. 5336 EMD type GP38-2 (built as Conrail No. 8168 in June 1977) is switching in the Binghamton (Conklin), New York, yard on December 1, 2018 painted with Operation Life Saver graphics. (*Brian A. Keates photograph*)

On April 17, 2017, NS Abrams Yard in King of Prussia, Pennsylvania, is the location of NS diesel locomotive No. 5642. This was originally Penn Central diesel locomotive No. 7868 (EMD type GP38 built in January 1971). It later became Conrail No. 7868 and still later NS No. 2923. It was upgraded to an EMD GP38-2 at the NS Juniata, Pennsylvania, shop in May 2006 and renumbered No. 5642. In November 2015, the locomotive received a special "Training First Responders" paint scheme. (*Brian A. Keates photograph*)

Leading on a westbound intermodal train through Cove, Pennsylvania, is NS diesel locomotive No. 5670 on February 17, 2018. This originally was Norfolk & Western No. 4100 type GP38AC. It was rebuilt to an EMD GP38-2 at the NS Juniata, Pennsylvania, shop in May 2007, renumbered No. 5670, and has been painted with Operation Life Saver graphics. (*Brian A. Keates photograph*)

On August 23, 2017, NS 4,400-horsepower diesel electric locomotive No. 9250 (GE type Dash 9-44CW placed in service on January 19, 1999 and painted in the twenty-fifth anniversary Operation Life Saver paint scheme) leads an eastbound general manifest train through Altoona, Pennsylvania. (*Brian A. Keates photograph*)

Westbound general manifest train passes through downtown Altoona, Pennsylvania, on August 24, 2019 headed by NS diesel electric locomotive No. 9254 (GE type Dash 9-44CW placed in service February 3, 1999 and painted in the twenty-fifth anniversary Operation Life Saver paint scheme). When more stringent emission regulations went into effect on January 1, 2005 in the United States, the Dash 9-44CW was replaced in production by the GE ES44DC. (*Brian A. Keates photograph*)

Rolling eastbound through Cresson, Pennsylvania, on August 21, 2018 is NS track test unit No. 34 (rebuilt from Norfolk & Western 2,500-horsepower six-axle diesel electric locomotive EMD type SD35 built in October 1965) which has measuring equipment for track geometry and rail wear followed by a car that processes data supplied by No. 34. (*Brian A. Keates photograph*)

In March 2016, NS locomotive No. 911 is eastbound through Fitchburg, Massachusetts, with an autorack train heading for Ayer, Massachusetts. This locomotive was originally NS EMD type SD60 No. 6577 built in July 1985. It was rebuilt by NS on April 30, 2015 and painted in a special red, white, gold, and black paint scheme "Honoring First Responders." (*Robert A. Elder photograph*)

3

Run-Through Power on Norfolk Southern Railway

Railroads share locomotives in an agreement known as a run-through power agreement or pooled power agreement. A train originates on one railroad and is handed over to another railroad to get it to its destination. Another reason is that one railroad may owe another railroad for the use of their power and is paid back with reciprocal use of power instead of money. As an example, at a Norfolk Southern Railway grade crossing, the train may pass by with Burlington Northern Santa Fe Railway locomotives. Railroads that have run-through agreements use whatever locomotives they have in their railroad yard. Railroad A receives an EMD SD40-2 diesel rated 3,000 horsepower from railroad B on a run-through and operates it for fifty hours before returning it to railroad B. Railroad A multiplies 3,000 horsepower of the SD40-2 locomotive by fifty hours which equals 150,000 horsepower hours. Hence railroad A owes horsepower hours to railroad B which would be paid back by lending their own locomotives to railroad B for an equivalent horsepower hour period of time.

Another reason for a different railroad company locomotive running could be a lease of that locomotive, which can occur when a surge of railroad traffic requires more locomotives than usual, when more locomotives than usual are suspended from operation due to inspections or maintenance reasons, or when a railroad has a locomotive unexpectedly out of service. A railroad that has excess power will lease locomotives to a railroad that is power short. Also, there could be a detour where one railroad's line gets washed out and has to detour over another railroad company route.

On August 4, 2022, Canadian Pacific 4,400-horsepower diesel electric locomotive No. 8138 is trailing on an eastbound NS intermodal train through Wyomissing, Pennsylvania. This locomotive was originally No. 9671 GE AC4400CW built in January 1998 rebuilt by GE into designation AC4400CWM with upgraded front cab, upgraded electrical systems, 10 percent fuel efficiency improvement, 40 percent increase in reliability, and a 50 percent increase in haulage ability. (*Brian A. Keates photograph*)

Kansas City Southern Railway (KCS) six-axle 4,300-horsepower diesel locomotive No. 4167 (type EMD SD70 ACe one of ten [Nos. 4165–4174] built during March to April 2014) is trailing on a military flat car train through Wyomissing, Pennsylvania, on March 8, 2022. The KCS (linking Kansas City, Missouri, with key ports along the Gulf of Mexico in the states of Alabama, Louisiana, Mississippi, and Texas) connects with the Kansas City Southern de Mexico (KCSM) which handles freight into northeastern and central Mexico plus to the Gulf of Mexico ports of Tampico, Altamira, and Vera Cruz. (*Brian A. Keates photograph*)

On January 19, 2022, a group of diesel locomotives headed by 4,400-horsepower NS No. 4196 (type AC44C6M rebuilt by Wabtec/GE Erie, PA, in June 2019 from 4,400-horsepower No. 8985 GE Dash 9-44CW that was originally built in April 1996) with trailing 4,000-horsepower locomotive CN No. 8817 (EMD type SD70M-2 built in 2007) are handling a westbound empty tank train through Cresson, Pennsylvania. (*Brian A. Keates photograph*)

NS local train K81 with trailing CSX Transportation Railroad (CSX) diesel locomotive No. 889 (GE type ES44AC-H built in April 2008) is heading south for Taylor yard on April 8, 2018. (*Brian A. Keates photograph*)

June 23, 2018, Ferromex 4,300-horsepower six-axle EMD type SD70ACe diesel locomotive No. 4079 (one of thirty locomotives [Nos. 4059–4088] built during September 2011 through December 2011 by Bombardier in Mexico) is trailing on a westbound NS intermodal train through Cove, Pennsylvania. The 5,970-mile Ferromex system began operation on February 19, 1990 in Mexico connecting five major inland Mexican cities, five cities along the border with the United States, four seaports on the Pacific Ocean, and one on the Gulf of Mexico. (*Brian A. Keates photograph*)

Florida East Coast Railway diesel locomotive No. 106 (EMD type SD70M-2 built in August 2008) is running on NS local train K81 out of Topton, a borough in Berks County, Pennsylvania, on August 10, 2018. (*Brian A. Keates photograph*)

On August 22, 2018, diesel locomotive CEFX No. 1050 (GE type AC44CW built in November 2004) is trailing on an NS eastbound intermodal train through Horseshoe Curve in Blair County, Pennsylvania. CIT Group, under the reporting mark CEFX, leases locomotives and railroad cars to rail transport and shipping companies in North America. (*Brian A. Keates photograph*)

Kansas City Southern Railway 4,400-horsepower diesel locomotive No. 4537 (GE type AC44CW originally Transportacion Ferroviaria Mexicana [TFM] built in May 1998) is rolling eastbound along the NS line through Wyomissing, Pennsylvania, on May 8, 2022 with a United States Department of Defense trailer train. (*Brian A. Keates photograph*)

On September 16, 2018, Progress Rail Services Corporation, reporting mark PRLX, locomotive No. 9553 (EMD type SD70MAC built in November 1994) is trailing on a westbound NS intermodal train through Cove, Pennsylvania. Progress Rail Services Corporation sells and leases locomotives plus is one of the largest suppliers of locomotives and rolling stock for the global rail industry. (*Brian A. Keates photograph*)

On November 30, 2021,Union Pacific Railroad diesel locomotive No. 8151 (GE type ES44AC-H built in May 2014) is trailing on an NS westbound intermodal train through Cresson, Pennsylvania. Cresson has served as a meeting point for the authors of this book to exchange ideas while waiting to photograph the next train. (*Brian A. Keates photograph*)

On Sunday August 21, 2022, Union Pacific Railroad (UP) diesel locomotive No. 8906 (EMD type SD70AH built in October 2014) is leading an NS westbound intermodal train at Wyomissing, Pennsylvania. This locomotive has an isolated "Whisper Cab" design to reduce noise and vibration to the crew and equipment inside the operator's cab. The UP's 281 locomotives of this type (Nos. 8824–9104) each have 8,000 pounds of extra ballast for more tractive effort. (*Brian A. Keates photograph*)

Westbound general manifest NS train lead by 4,400-horsepower diesel electric Burlington Northern Santa Fe Railway (BNSF) locomotive No. 4365 (GE type C44-9W built in March 1999) is about to cross Walbridge Road in Harborcreek Township, in Erie County, Pennsylvania, on October 10, 2022. Just north of this NS line is the CSX Erie West Subdivision. (*Kenneth C. Springirth photograph*)

On November 4, 2022, westbound train headed by Union Pacific (UP) diesel locomotive No. 5706 (GE type AC4400CW with Controlled Tractive Effort software making the designation AC44CW-CTE built in March 2001) trailed by BNSF diesel locomotive No. 5401 (GE type C44-9W built in September 2000) is nearing the Walbridge Road crossing in Harborcreek Township, Erie County, Pennsylvania. (*Kenneth C. Springirth photograph*)

Walbridge Road in Harborcreek Township, Erie County, Pennsylvania, is the location on October 12, 2022 of westbound 4,400-horsepower NS locomotive No. 9911 (GE type D9-44CW placed in service on October 13, 2004) and 4,400-horsepower Canadian National Railway locomotive No. 2786 (GE type ES44AC originally Citicorp Railmark Inc. [CREX] No. 1401) delivered during the summer of 2014 and later sold to CNR. (*Kenneth C. Springirth photograph*)

On November 12, 2022, a southbound NS general merchandise train is preparing to go under the Paul E. Sweeney Memorial Bridge in the borough of Lilly in Cambria County, Pennsylvania, headed by NS locomotive No. 8034 (GE type ES44AC placed in service on November 21, 2010) and trailing UP locomotive No. 4139 (EMD type SD70M built in September 2000). (*Kenneth C. Springirth photograph*)

4

Horseshoe Curve & Altoona

In 1800, a trip from Philadelphia to Pittsburgh on horseback required a month. The trip time was cut to 4.5 days on March 18, 1834 by completion of the state-owned Main Line of Public Works which consisted of the Philadelphia & Columbia Railroad (Philadelphia to Columbia), Eastern Division Canal (Columbia to Duncan's Island at the mouth of the Juniata River), Juniata Division Canal (Duncan's Island to Holidaysburg), Allegheny Portage Railroad (Holidaysburg to Johnstown), and the Western Division Canal (Johnstown to Pittsburgh).

The Pennsylvania Railroad was incorporated on April 13, 1846 to build a railroad from Harrisburg to Pittsburgh. John Edgar Thomson, chief engineer for the Pennsylvania Railroad, chose a route along the flatter terrain along the Juniata River until the foothills of the Allegheny Mountains where the line climbed a moderate grade for 10 miles where it reached the split of two mountain ravines which were crossed where about 450 Irish labors, famous for their strong work ethic, with picks and shovels built the fill (all done without any heavy equipment) with the tracks ascending a 220-degree curve, about 1,800 feet across at its widest point, and 2,375 feet in length. The north side has a 637-foot radius, and the south side has a 609-foot radius. East side of the curve is 122 feet lower than the west side of the curve. The grade of the curve is 91 feet per mile, and the east end elevation is 1,594 feet above sea level. It became known as the Horseshoe Curve. The crest of the mountain was penetrated by the Gallitzin Tunnels from which the line descended by a more moderate grade to Johnstown. The 31.1-mile line between Altoona and Johnstown including the Horseshoe Curve opened on February 15, 1854. With the opening of the Harrisburg to Pittsburgh line, travel time from Philadelphia to Pittsburgh ranged from thirteen to seventeen hours depending on which train passengers took. By the end of 1854, Horseshoe Curve became a double-track line. As demand for train travel increased, a third track was added in 1898 and a fourth track was added in 1900.

Five miles west of Altoona, in Logan Township, Blair County, Pennsylvania, Horseshoe Curve is at milepost 242 on the Pittsburgh Line (connecting Harrisburg with Pittsburgh) on the Norfolk Southern Railway. Westbound trains climb a maximum grade of 1.85 percent for 12 miles from Altoona to Gallitzin. Slightly west of the Gallitzin Tunnels, trains reach the summit of the Allegheny Mountains and then descend for 25 miles at a maximum grade of 1.1 percent.

During World War II, Horseshoe Curve transported troops plus material for the Allied war effort and was placed under armed guard. Germany's military intelligence planned to sabotage important industrial assets in the United States. In June 1942, four men were brought by submarine landing on the coast of Long Island. They were assigned to destroy the Horseshoe Curve, Hell Gate Bridge, Alcoa aluminum factories, and locks on the Ohio River. Saboteur George John Dasch turned himself in, and the rest of the saboteurs were promptly captured by the Federal Bureau of Investigation.

Pennsylvania Railroad steam locomotive No. 1361 was placed on display at the Horseshoe Curve park on June 8, 1957 and later removed to permit its restoration to operating condition. In its place, EMD type GP9 diesel No. 7048 was positioned at the Horseshoe Curve on September 16, 1985. The Horseshoe Curve was designated a National Historic Landmark on November 13, 1966. Under Consolidated Rail Corporation (Conrail), cuts were made in the system that resulted in reducing the number of tracks at the Horseshoe Curve from four to three.

In 2004, the Horseshoe Curve was dedicated as a National Historic Civil Engineering Landmark that noted:

> One of the earliest and most impressive of America's great engineering feats, the Horseshoe Curve was built as a means of overcoming a straight-line grade over the geological feature known as the Allegheny Escarpment or Allegheny front, which separates the ridge-and-valley section of Pennsylvania (on the east) from the Allegheny Front (on the west). Such a straight-line route would have made commercial railroad operations unfeasible from both an economic and technical standpoint.

On July 23, 1998, the Federal Surface Transportation Board approved the plan for Norfolk Southern to buy 58 percent of Conrail including the former Pennsylvania Railroad main line over the Horseshoe Curve and CSX 42 percent of Conrail which became effective on June 1, 1999.

Five miles east of Horseshoe Curve is the city of Altoona, Pennsylvania. Located in downtown Altoona behind the Altoona Railroad Memorial Museum is the "Railwalk" built for train watching. This is a sidewalk along the railroad tracks protected by a 4-foot-high chain-link fence. On June 16, 2012, Norfolk Southern's Alto Tower closed completing a project upgrading interlockings within the 5-mile Altoona terminal.

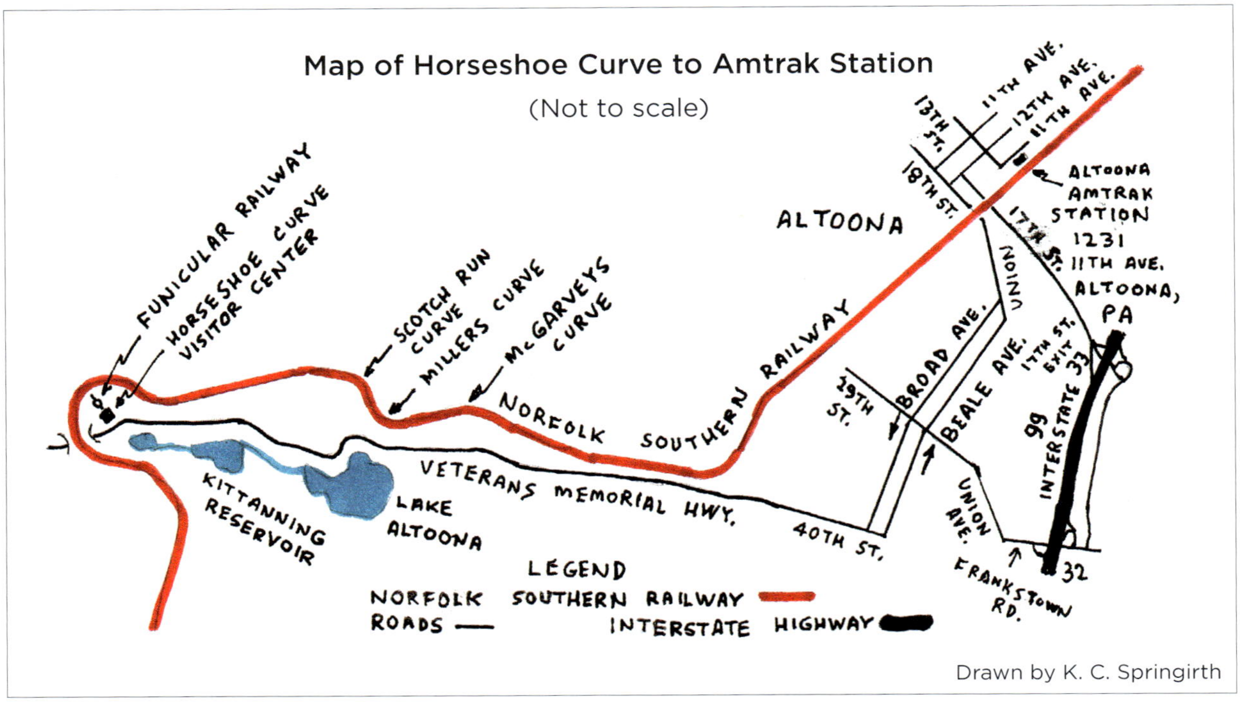

Opening on February 15, 1854, the well-designed and built Horseshoe Curve has been in continuous operation. This monumental project was the work of immigrants by manual effort with picks, shovels, and dynamite to cut out portions of the mountain. Horseshoe Curve is accessible to downtown Altoona and Interstate 99 as shown in the above map.

On June 19, 2022, co-author of this book Kenneth C. Springirth is at Horseshoe Curve alongside Pennsylvania Railroad (PRR) four-axle 1,750-horsepower diesel locomotive No. 7048 (EMD type General Purpose [GP9] built in December 1955). This locomotive was placed on display at the Horseshoe Curve on September 16, 1985. The PRR had 270 GP9 locomotives (Nos. 7000–7269). On the Pennsylvania Railroad, a GP9 was classed as an EFS-17m with E for Electro Motive as builder, FS for Freight Shifter, 17 for the locomotive's 1,750-horsepower rating, and m for multiple unit-control. (*Virginia M. Springirth photograph*)

Eastbound NS diesel locomotives Nos. 4311(originally NS No. 9227 GE Dash 9-44CW built May 1998) rebuilt to type AC44C6M by Wabtec/GE at Ft. Worth, Texas, in December 2020, No. 4232 (originally NS No. 9106 GE Dash 9-44CW built in April 1997) rebuilt to type AC44C6M at NS Roanoke Locomotive Shop in June 2019, and No. 1061 (type EMD SD70ACe placed in service on December 23, 2011) are passing the Horseshoe Curve on June 19, 2022. (*Kenneth C. Springirth photograph*)

On October 26, 2022, NS No. 4183 (originally NS No. 8952 GE Dash 9-44CW built in March 1996 rebuilt to type AC44C6M by Wabtec/GE at Erie, PA in March 2019) leads a westbound intermodal train through the Horseshoe Curve west of Altoona, Pennsylvania. (*Brian A. Keates photograph*)

Surrounded by the beautiful October 6, 2022 fall foliage, NS No. 4239 (originally NS No. 9024 GE Dash 9-44CW built in February 1997 rebuilt to type AC44C6M at NS Juniata Locomotive Shop in December 2019) heads an empty coal train through the Horseshoe Curve. (*Brian A. Keates photograph*)

Amid the magnificent change in fall foliage, 4,400-horsepower NS locomotive No. 3667 (GE type ET44AC placed in service on March 3, 2017) is leading a westbound intermodal train over the Horseshoe Curve on October 6, 2022. (*Brian A. Keates photograph*)

A short utility train is powered by 4,400-horsepower NS locomotive No. 4296 (originally NS No. 9092 GE Dash 9-44CW built in April 1997 rebuilt to type AC44C6M at the NS Juniata Locomotive Shop in June 2020) eastbound on the Horseshoe Curve on October 6, 2022. The AC44C6M replaces the Dash 9's DC traction motors with alternating current GE motors. (*Brian A. Keates photograph*)

NS locomotive No. 9747 (GE type D9-44CW placed in service on January 29, 2003) leads an eastbound train passing a westbound tank train on the Horseshoe Curve on October 6, 2022. (*Brian A. Keates photograph*)

On October 6, 2022, two NS locomotives No. 7245 type SD70ACU (originally UP No. 8016 built in February 1996, renumbered UP No. 3488, and rebuilt from NS SD90MAC No. 7245 by NS in June 2016) and No. 7253 type SD70ACU (originally UP No. 8024 built in March 1996, renumbered UP No. 3492, and rebuilt from NS SD90MAC No. 7253 in October 2016) are trailing on an eastbound intermodal train passing over the Horseshoe Curve. (*Brian A. Keates photograph*)

Powered by Amtrak locomotive No. 160 (type GE P42DC [4,250 horsepower at 1,047 rpm and maximum speed of 110 mile per hour] built in April 2001), the eastbound *Pennsylvanian* Amtrak passenger train is at the Altoona Transportation Center 1231 11th Avenue in Altoona, Pennsylvania, on May 14, 2013. Connecting New York City via Philadelphia to Pittsburgh, this Amtrak passenger train is the only passenger train operating across Pennsylvania. (*Kenneth C. Springirth photograph*)

On March 15, 2013, westbound NS 4,300-horsepower locomotive No. 1032 (EMD type SD70ACe placed in service on November 4, 2011) is leading a string of empty coal cars as seen from the 12th Street foot bridge in downtown, Altoona, Pennsylvania. (*Kenneth C. Springirth photograph*)

On October 4, 2022, NS locomotives No. 754 EMD/NS type RP-E4C Slug (built on a newly constructed frame by NS in February 2015) and No. 3011 EMD type GP40-2 (originally Conrail No. 3294 built in May 1977 and used by NS as a slug master for RP-4EC slugs) are passing the Altoona Transportation Center returning to Rose Yard, a former classification yard operated by the Pennsylvania Railroad and later Conrail. The classification yard has been decommissioned and part of the yard is still used for switching. (*Brian A. Keates photograph*)

On June 27, 2022, NS 4,400-horsepower locomotive No. 7574 (GE type ES44DC placed in service on June 22, 2006) leads an eastbound general manifest train by the Altoona Transportation Center. (*Brian A. Keates photograph*)

Westbound NS locomotive No. 4026 (originally NS No. 8875 GE Dash 9-40C built in March 1995 rebuilt to type AC44C6M at GE Ft. Worth, Texas, in May 2017) leads an empty coal train through downtown Altoona, Pennsylvania, near the Altoona Transportation Center on October 8, 2022. (*Brian A. Keates photograph*)

On November 12, 2022, NS locomotive No. 8172 (GE type ES44AC placed in service on April 20, 2017) is rolling through downtown Altoona, Pennsylvania, heading a westbound general manifest train. (*Brian A. Keates photograph*)

Rolling past an NS track crew truck at Brickyard Crossing in Altoona, Pennsylvania, on October 2, 2022 is eastbound locomotive NS No. 4536 (originally No. 9290 GE Dash 9-44CW built in April 1999 rebuilt to type AC44C6M at Wabtec/GE Ft. Worth, Texas, in May 2022). (*Brian A. Keates photograph*)

On November 12, 2022, NS helper locomotive No. 7254 type SD70ACU (originally UP No. 8035 built in June 1996, renumbered UP No. 3518, and rebuilt from NS SD90MAC No. 7254 by Progress Rail in December 2016) is trailing another helper locomotive pushing a westbound intermodal train out of Altoona, Pennsylvania, passing Alto Tower. Built in 1915, a re-signaling project for the Altoona terminal resulted in the closing of Alto Tower in 2012. (*Brian A. Keates photograph*)

This is a November 7, 2020 view of the NS Juniata shop showing numerous locomotives and a neat turntable. According to Norfolk Southern, this shop can strip a locomotive down to its bare frame and completely rebuild it in 6.5 days. In addition, this shop has its own locomotive emissions testing facility. (*Brian A. Keates photograph*)

5

NS in Gallitzin, Cresson, Lilly, and Cassandra, Pennsylvania

Gallitzin Borough has the active Allegheny Tunnel and north of it the parallel Gallitzin Tunnel that was closed in 1995. Both of these tunnels can be viewed from the Jackson Street bridge. South of these tunnels and just north of Portage Street is the active New Portage Tunnel. Gallitzin has lost 57.5 percent of its population declining from 3,618 in 1940 to 1,536 in 2020.

The borough of Cresson in Cambria County, Pennsylvania, has a railroad observation platform located on the east side of the Norfolk Southern Railway tracks. Cresson has lost 42.6 percent of its population, declining from 2,659 in 1960 to 1,525 in 2020. The July 12, 2021 Altoona Mirror newspaper noted "Norfolk Southern furloughed 86 workers at its Juniata Locomotive Shop in Altoona on Friday while eliminating another 14 jobs through attrition." On July 21, 2021, the Altoona Mirror reported that "Norfolk Southern will close its engine house in Cresson, PA eliminating seven jobs." That building (still there in July 2022) was across the tracks from the Railroad Observation Platform.

South from Cresson on Highway 53 is Carneys Crossing, a grade crossing over the Norfolk Southern Railway main line, and further south is Lilly, a borough in Cambria County, Pennsylvania. Lilly has lost 62.5 percent of its population, declining from 2,346 in 1920 to 879 in 2020. On October 12, 2018, the Paul E. Sweeney Memorial Bridge (an excellent site to photograph trains) was dedicated on Overbridge just west of Railroad Street in Lilly. Born in 1919 in Lilly, Sweeney entered the army at age twenty-two, trained as a combat engineer, and saw action in the Saint Lô campaign after landing at Normandy. He also received the Purple Heart for injuries he received from a German sniper.

South of Lilly on Highway 53 is the borough of Cassandra where there is an iron bridge that serves as a train overlook. Cassandra has lost 73.5 percent of its population, from 514 in 1940 to 136 in 2000. Fortunately, there has been a slight improvement to 162 in 2020.

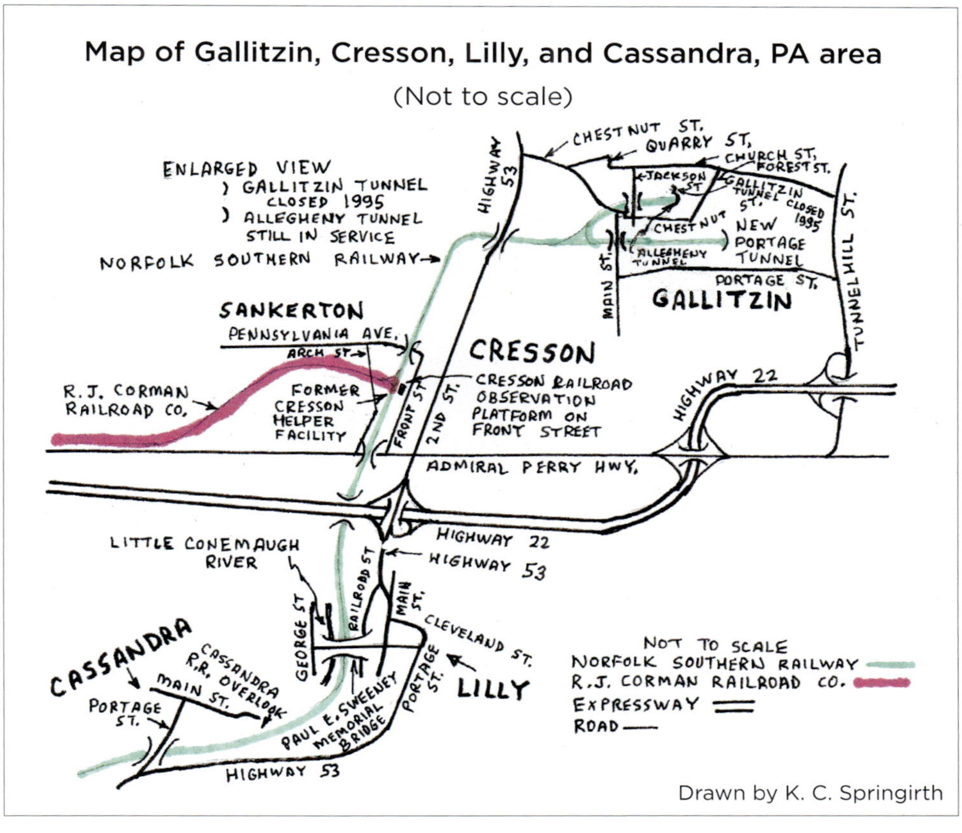

The above map shows rail viewing sites in Gallitzin, Cresson, Lilly, and Cassandra. It is important to keep off railroad property.

On a snow-covered February 18, 2021, a westbound NS locomotive type SD70ACU No. 7247 (originally UP No. 8018 built in February 1996, renumbered UP No. 3489, and rebuilt from NS SD90MAC No. 7247 by NS in April 2017) has emerged from the Allegheny Tunnel in Gallitzin, Pennsylvania. Overlooking the site is former Pennsylvania Railroad PRR class N5C caboose (AAR type M930) No. 477852 built in 1942. (*Brian A. Keates photograph*)

On February 14, 2021, a westbound NS intermodal train is coming out of the Allegheny Tunnel at milepost 248 in Gallitzin, Pennsylvania, powered by NS locomotive No. 8074 (GE type ES44AC placed in service on February 1, 2011). (*Brian A. Keates photograph*)

On Saturday October 23, 2021, NS locomotive No. 9978 (GE type D9-44CW placed in service on December 20, 2004) is leading an eastbound intermodal train through Cresson, Pennsylvania, on track one. (*Brian A. Keates photograph*)

NS locomotive No. 3662 (GE type ET44AC placed in service on March 2, 2017) leads a westbound intermodal train through Cresson, Pennsylvania, with the third locomotive being a trailing Union Pacific locomotive on Saturday December 4, 2021. (*Brian A. Keates photograph*)

On May 24, 2020, two NS EMD type SD40E helper locomotives No. 6314 (originally Conrail No. 6754 EMD type SD50 built in May 1984, became NS No. 5431, and rebuilt to NS No. 6314 EMD SD40E at NS's Juniata shops on June 30, 2009) and No. 6306 (originally NS No. 6308 EMD SD50 built in 1984 and rebuilt to an EMD SD40E on February 11, 2009) are rolling eastbound through Cresson, Pennsylvania. (*Brian A. Keates photograph*)

On July 19, 2022, NS locomotive No. 4502 (originally NS No. 9282 GE Dash 9-44CW built in February 1999 and rebuilt to type AC44C6M at NS Juniata Locomotive Shop in June 2022) is leading a fully loaded coal train eastbound through Cresson, Pennsylvania. (*Brian A. Keates photograph*)

NS locomotive No. 4396 (originally No. 9170 NS GE Dash 9-44CW built in March 1998 and rebuilt to type AC44C6M at Wabtec/GE Ft. Worth, Texas, in September 2020) and trailing NS locomotive No. 4002 (originally NS No. 8789 GE Dash 9-40C built in January 1995 and rebuilt to type AC44C6M by NS Roanoke shop in August 2016 in a special paint scheme with DC to AC lettering and AC waveform under the cab numbers) lead a general manifest train through Cresson, Pennsylvania, on October 6, 2022. (*Brian A. Keates photograph*)

On June 19, 2022, the morning eastbound Amtrak *Pennsylvanian* is passing through Cresson, Pennsylvania, via the Pittsburgh Line, NS's busiest freight corridor, powered by GE type P42DC locomotive No. 117 that was built in September 1997. P42 DC locomotives are primarily used on Amtrak's long haul passenger service. (*Brian A. Keates photograph*)

An eastbound NS intermodal train is traveling through Cresson, Pennsylvania, on June 18, 2022 powered by NS locomotive No. 4362 (originally NS No. 9087 GE Dash 9-44CW built in March 1997 and rebuilt to type AC44C6M in July 2020 by Wabtec/ GE at Ft. Worth, Texas) with a Burlington Northern Santa Fe Railway diesel behind it. (*Brian A. Keates photograph*)

On Sunday August 26, 2018, NS locomotive No. 7209 EMD type SD80MAC (originally Conrail locomotive No. 4117 EMD type SD80MAC built in March 1996) is at the NS locomotive service center in Cresson, Pennsylvania, which closed in July 2021. (*Brian A. Keates photograph*)

NS locomotive No. 7249 type SD70ACU (originally UP No. 8020 built in February 1996, renumbered UP No. 3475, and rebuilt from NS No. 7249 type SD90MAC in December 2016 by NS) is the last of two helper locomotives on the rear of an eastbound general manifest train traveling through Cresson, Pennsylvania, on November 2, 2021. (*Brian A. Keates photograph*)

Just after a winter snowstorm on February 16, 2022, NS locomotive No. 1176 (one of sixty type SD70ACe -T4C locomotives [Nos. 1175–1234] built in 2018–2019 by Progress Rail to meet tier 4 regulations) and NS No. 9813 (GE type D9-44CW placed in service on March 15, 2004) are leading a westbound intermodal train at Cresson, Pennsylvania. (*Brian A. Keates photograph*)

On October 5, 2022, diesel locomotive No. 7117 (3,000-horsepower EMD SD40-2 built in December 1978 for the Burlington Northern Railroad) is one of three RJ Corman Railroad Company locomotives ready to hook to a grain train at Cresson, Pennsylvania. This is a connection point with the Norfolk Southern Railway. (*Brian A. Keates photograph*)

On October 6, 2022, Union Pacific (UP) locomotive No. 8088 is trailing on a westbound NS general manifest train passing under the Paul E. Sweeney Memorial Bridge in Lilly, Pennsylvania. Sweeney saw action in the Saint Lô campaign and received a silver star for his combat actions and was awarded the Purple Heart for injuries he received from a German sniper. (*Brian A. Keates photograph*)

Eastbound Amtrak passenger diesel locomotive No. 87 (GE type P42DC built in May 1997) is under the Paul E. Sweeney bridge at Lilly, Pennsylvania, on October 6, 2022. Sweeney received the Silver Star and Purple Heart in the United States plus the British Military Medal inscribed by the King of England for his service. After returning to the United States, Sweeney worked as a train crewman for the Pennsylvania Railroad and later Conrail. (*Brian A. Keates photograph*)

On October 6, 2022, NS locomotive No. 4515 (originally NS No. 9390 GE Dash 9-44CW built in June 1999, rebuilt to type AC44C6M at NS Juniata Locomotive Shop in September 2022) is leading an empty westbound coal train near the Paul E. Sweeney bridge at Lilly, Pennsylvania. (*Brian A. Keates photograph*)

Amid the brilliant autumn foliage of October 5, 2022, NS locomotive No. 4003 (originally NS No. 8839 GE Dash 9-40C built in February 1995, rebuilt to type AC44C6M at NS Roanoke Shop on October 16, 2019 and painted in a special black, Tuscan red, gray, and white paint scheme with DC to AC lettering and AC waveform under the cab numbers) is westbound at Carney's Crossing at Lilly, Pennsylvania. (*Brian A. Keates photograph*)

On November 8, 2022, Carney's Crossing is the location of a westbound intermodal train headed by NS locomotive No. 7710 (GE type ES44DC placed in service on February 12, 2008). (*Brian A. Keates photograph*)

Two NS helper locomotives No. 7236 type SD70ACU (originally UP No. 8006 built in January 1996, renumbered UP No. 3476, and rebuilt from NS SD90MAC No. 7236 by NS in November 2016) and No. 7253 type SD70ACU (originally UP No. 8024 built in March 1996, renumbered UP No. 3492, and rebuilt from NS SD90MAC No. 7253 in October 2016) are rolling eastbound at Carney's Crossing at Lilly, Pennsylvania, on December 18, 2021. (*Brian A. Keates photograph*)

On November 8, 2020, NS locomotive No. 4190 (originally NS No. 9000 GE Dash 9-44CW built in April 1996 rebuilt to type AC44C6M at Wabtec/GE-Erie, PA, in May 2019) is leading an eastbound grain train near Carney's Crossing. (*Brian A. Keates photograph*)

The scenic fall foliage at Cassandra, Pennsylvania, on October 6, 2022, is the location of NS locomotive No. 9542 (GE type D9-44CW placed in service on December 18, 2000) leading a westbound intermodal train that is passing under the pedestrian bridge known as the Cassandra Overlook. (*Brian A. Keates photograph*)

On October 6, 2022, Cassandra, Pennsylvania, is the location of NS helper locomotive No. 7253 type SD70ACU (originally UP No. 8024 built in March 1996, renumbered UP No. 3492, and rebuilt from NS SD90MAC No. 7253 in October 2016) along with another helper are rolling on the rear of a westbound intermodal train. (*Brian A. Keates photograph*)

6

Harris Tower Railroad Museum

Harris Tower, located at 7th and Walnut Streets in Harrisburg, Pennsylvania, was constructed by the Pennsylvania Railroad and opened on April 26, 1930. Harris Tower was the western terminus of the Pennsylvania Railroad's electrified passenger lines. At Harris Tower, westbound trains had their electric locomotives removed and replaced by steam or diesel. For eastbound trains, the process was reversed. Eastbound trains were split into a New York and Philadelphia section and a Baltimore and Washington section. For westbound trains, the process was reversed. The process also involved adding or removing cars. The tower had 113 interlocking levers that controlled switches and signals along 3,250 feet of track from Herr Street to the middle of the Harrisburg passenger station platforms. The Pennsylvania Railroad operated the tower from 1930 to 1968, followed by Penn Central Transportation Company from 1968 to 1976, and Amtrak from 1976 to 1992.

The completion of the Interstate Highway System and more people flying resulted in a dramatic decline in rail passenger travel. Fewer trains resulted in a reduction of track in Harrisburg, plus with computerized and centralized regional control operations, Harris Tower declined in importance and was closed on November 15, 1991. The Harrisburg Chapter of the National Railway Historical Society (NRHS) purchased Harris Tower on July 24, 1992 to be preserved as a publicly accessible museum. Harrisburg Chapter's efforts resulted in having the building listed on the National Register of Historic Places on August 30, 1994. Visitors to the museum learn about the tower's former operations by means of a computer program specially designed for use at Harris Tower. Connected to the original 113 interlocking levers, visitors are able to control virtual trains over a simulation of the 1940s track pattern using Pennsylvania Railroad timetables and train schedules from that time period. The tower is popular as an observation point for the trains that pass it.

In this August 6, 2022 view, Harris Tower is shown at 637 Walnut Street in Harrisburg, Pennsylvania. The purpose of the tower was to control the former labyrinth of tracks at the north (railroad west) end of the Harrisburg Passenger Station. The territory served by the tower in 1945 had eighty-seven switches and 106 signals; in 1982, it had eighteen switches and twenty-nine signals; and in 1991, it had four switches and ten signals according to the National Railway Historical Society "Harris Interlocking Tower Fact Sheet." (*Brian A. Keates photograph*)

On August 20, 2022, NS locomotive No. 4335 (originally NS No. 9043 GE Dash 9-44 CW built in February 1997, rebuilt to type AC44C6M at Wabtec/GE-Ft. Worth, Texas, in April 2020) and NS locomotive No. 4282 (originally NS No. 9105 GE Dash 9-44CW built in April 1997, rebuilt to type AC44C6M at NS Juniata Locomotive Shop in April 2020) are passing Harris Tower in Harrisburg while leading an eastbound intermodal train. (*Brian A. Keates photograph*)

On May 10, 2022, Norfolk Southern Railway (NS) train 39G is westbound towards the Rockville Bridge in Harrisburg, Pennsylvania, pulled by five locomotives including CSX No. 4406 (EMD type GP40-2 originally built in May 1977 as No. 3295 for Conrail). There were 129 GP40-2 locomotives (Nos. 3275–3403) built for Conrail, of which fifty-three went to CSX becoming Nos. 4400–4452 and seventy-one went to NS becoming 3000–3070. (*Brian A. Keates photograph*)

On August 20, 2022, Amtrak GE type P42DC (officially trademarked Genesis) passenger diesel locomotive No. 116 (built in September 1997) is leading a westbound passenger train out of the Harrisburg, Pennsylvania, station past the Harris Tower. The P42DC is primarily used on Amtrak's longer haul and higher speed service. (*Brian A. Keates photograph*)

A bright cloudy August 6, 2022 finds Amtrak P42DC diesel locomotive No. 125 (built in October 2000) leading an eastbound passenger train to the Harrisburg, Pennsylvania, train station. This 69-foot-long locomotive has a 10-foot width, 14.33-foot height, and weighs 268,240 pounds. (*Brian A. Keates photograph*)

On May 10, 2022, Union Pacific Railroad (UP) locomotive No. 5515 (GE type ES44AC built in June 2005) is leading an eastbound NS intermodal train on NS tracks past Harris Tower in Harrisburg, Pennsylvania. This locomotive is UP class C45ACCTE because UP uses CTE software to reduce the tractive effort of the locomotive so it can be used as a distributed power unit (DPU). A DPU is a locomotive or locomotives that can be placed anywhere in a train and be controlled by the leading locomotive which has greatly increased the length of a train. (*Brian A. Keates photograph*)

On January 27, 2018, Amtrak electric locomotive No. 642 (Siemens Mobility type ACS-64, one of nineteen locomotives of this type [Nos. 631–649] delivered in 2015) is at the Amtrak Harrisburg, Pennsylvania, station. This locomotive (designed to run at speeds of 125 miles per hour) is Siemen Mobility's first transfer of a high-speed locomotive from Europe to the United States. Honoring all those who have served in the armed forces, the locomotive has a special "America's Railroad Salutes Our Veterans" color scheme painted at the Amtrak Beech Grove, Indiana, maintenance facility. (*Brian A. Keates photograph*)

7

Norfolk Southern Sunbury Line

The Norfolk Southern Sunbury Line operates from Sunbury, Pennsylvania (where it connects with Norfolk Southern's Buffalo Line), via Taylor Yard (in the borough of Taylor near Scranton in Lackawanna County, Pennsylvania) to Binghamton, New York (where it connects with Norfolk Southern's Southern Tier Line). Taylor Yard originally had twenty tracks and was and still is a staging yard of approximately ten tracks allowing freight cars to be stored for their next operation. This was once the Pennsylvania Railroad's Wilkes-Barre Branch, later part of the Delaware, Lackawanna & Western Railroad, in 1976 taken over by Conrail, later acquired by the Delaware & Hudson Railway (D&H) which was taken over by the Canadian Pacific Railway. On September 18, 2015, Norfolk Southern (NS) purchased the Sunbury–Binghamton line, renaming it Sunbury Line plus the D&H Binghamton to Schenectady line which NS labeled as its "Freight Line."

The September 18, 2015 Norfolk Southern News Release noted, "Norfolk Southern Corporation today completed the acquisition of 282 miles of the Delaware & Hudson Co.'s (D&H) line between Sunbury, Pa and Schenectady, NY, a $214.5 million transaction that increases options for rail carriers and supports jobs in Pennsylvania, New York, and New England." NS also noted, "The acquired lines connect with the NS network at Sunbury, PA and Binghamton, NY, and they provide NS single-line routes from Chicago and the southeastern United States to Albany, NY and NS intermodal terminals in Scranton, PA and Mechanicville, NY."

On August 18, 2017, NS locomotive No. 1126 (EMD type SD70ACe placed in service on January 3, 2015) with trailing Union Pacific Railroad (UP) locomotive No. 8562 (EMD type SD70ACe built in 2007) plus another NS locomotive are eastbound out of the Conklin/Binghamton area powering an intermodal train to Massachusetts. The young man taking a picture of the train is Jacob A. Keates, the son of Brian and co-author Beth Anne Keates. (*Brian A. Keates photograph*)

A beautiful summer day on July 30, 2022 finds NS locomotive No. 1024 (EMD type SD70ACe placed in service on April 11, 2011) switching freight cars at the Binghamton, New York, NS railway yard.
(*Brian A. Keates photograph*)

On a cloudy bright April 9, 2022, UP locomotive No. 7926 (GE ES44AC reclassified by UP as C45ACCTE built in May 2012) is handling a yard job at the Binghamton, New York, NS yard. (*Brian A. Keates photograph*)

NS locomotive No. 9811 (GE type D9-44CW placed in service on March 27, 2003) leads general manifest train No. 14R through Taylor, Pennsylvania, near Scranton on Saturday July 27, 2019. (*Brian A. Keates photograph*)

On May 5, 2018, former CSX locomotive No. 7361 (GE type C40-8W originally built as Conrail No. 6201 in July 1993) is switching cars at Taylor Yard setting up local train K81. This locomotive later became GECX (General Electric Company Transportation Systems Business Operations) No. 7361. An October 3, 2016 Wells Fargo news release noted Wells Fargo purchased "large components of GE Capital's rail platforms and commercial real estate businesses." (*Brian A. Keates photograph*)

Switching freight cars at Taylor Yard near Scranton, Pennsylvania, on August 10, 2018 is Reading Blue Mountain & Northern Railroad (RBMN) locomotive No. 3057 (EMD type SD40-2 originally built in January 1975 for the Chicago & Northwestern Railroad and later became Union Pacific No. 3066). The RBMN set a business record in 2017 handling 31,175 carloads which was a 50 percent increase in five years. (*Brian A. Keates photograph*)

On September 3, 2017, NS locomotive No. 5095 (EMD type GP38-2 built in October 1973) and trailing North Shore Railroad locomotive No. 1948 (EMD type SW1500 built in March 1970) are powering NS train No. 14R through Taylor Yard. (*Brian A. Keates photograph*)

On March 23, 2019, a northbound NS intermodal train is passing through Taylor, Pennsylvania, headed by NS locomotive No. 9480 (GE type D9-44CW placed in service on April 20, 2000). (*Brian A. Keates photograph*)

On July 30, 2022, New York Susquehanna & Western Railway (NYSW) No. 4060 (formerly Norfolk Southern Railway locomotive No. 2751 [EMD type SD70M-2 placed in service on May 11, 2006]) and NYSW locomotive No. 3012 (rebuilt EMD type SD33ECO [originally Southern Pacific Transportation Company locomotive No. 8242 type SD40T-2 built in April 1980]) are awaiting their next assignment at Binghamton, New York. (*Brian A. Keates photograph*)

On February 1, 2017, NS locomotive No. 3379 (EMD SD40-2 [originally Conrail No. 6445 built in September 1978]) is working at Taylor Yard as a local near Scranton, Pennsylvania. (*Brian A. Keates photograph*)

On December 1, 2018, NS 4,000-horsepower locomotive No. 7030 (originally 3,800-horsepower NS No. 6672 EMD SD60 built in December 1988 and rebuilt to SD60E by NS on May 15, 2016 uprating the engine to 4,000 horsepower and featuring a new NS wide nose cab) is leading the last Toys for Tots run up to Binghamton, New York. According to the November 26, 2019 *Daily Star* newspaper: "The annual Capital Region Toys for Tots train, which makes stops in Cobleskill, Oneonta, and Bainbridge, will not be running this year according to officials. Instead a convey of tractor trailers, buses and emergency vehicles will deliver the toys." (*Brian A. Keates photograph*)

8
Enola Yard and Harrisburg Yard

In 1902, the Pennsylvania Railroad (PRR) purchased 2 acres of land from farmers Francis and Wesley Miller for a station. Mr, Miller was given the honor of naming the station which he named after his daughter "Enola." As more people moved into the area, the community became known as Enola. In 1902, the PRR made the decision to build Enola Yard, which opened in 1905 located in East Pennsboro Township, Cumberland County, Pennsylvania, along the western side of the Susquehanna River across from Harrisburg, Pennsylvania. In the late 1920s, the 316-acre Enola Yard had a capacity of 9,692 cars. During 1939, the average daily train load was 11,207 cars rising to 14,100 daily cars for the World War II effort in October 1941. The westbound complex was rebuilt in 1944 with sixteen receiving tracks to hold 1,721 cars and thirty-five classification tracks, and the electrified eastbound complex had fifteen receiving tracks to hold 1,948 cars and thirty-three classification tracks. There was a hump yard for each that held 2,668 cars. By 1953, the yard had 145 miles of track and 476 switches. Conway Yard near Pittsburgh was rebuilt and after it reopened in 1959, became the PRR's major east west yard resulting in a decline in traffic at Enola Yard. After Conrail took over operation of Enola Yard in 1976, traffic declined with more trains routed over the nearby former Reading Railroad and Lehigh Valley Railroad lines. Conrail removed the electrified eastbound hump yard in 1983, closed the westbound hump yard in 1993, and closed the steel car shop in 1996. After Norfolk Southern Railway (NS) took over operations in 1999, operations have increased. By 2005, the yard operated seventy-nine tracks and handled 275,000 tons of freight daily. On the east side of the Susquehanna River is the NS Harrisburg Yard. It is important to remember to never trespass on railroad property.

As seen from the Train Watchers Bridge (River Road) south of Marysville, Pennsylvania, and north of the Interstate 81 bridge over the Susquehanna River, NS locomotive No. 6347 type SD40E (originally Conrail [CR] No. 6752 SD50 built in May 1984, and rebuilt from NS SD50 No. 5429 by NS in January 2012) is one of two locomotives setting up trains at the Enola Yard on the west side of the Susquehanna River on August 20, 2022. (*Brian A. Keates photograph*)

On August 20, 2022 NS locomotive No. 3625 (GE type ET44AC placed in service on June 1, 2016) is one of four locomotives heading out of Enola Yard northbound to pass under the Interstate 81 bridge shown ahead in the picture and later to cross the Rockville Bridge. (*Brian A. Keates photograph*)

Leading a general manifest train through the NS Harrisburg, Pennsylvania, yard on March 18, 2017, as seen at the end of Division Street near N. 7th Street in Harrisburg, Pennsylvania, is NS locomotive No. 7531 GE type ES44DC placed in service on December 28, 2005. (*Brian A. Keates photograph*)

9

Annville, Pennsylvania, Station

On January 18, 1858, the 55-mile Lebanon Valley Railroad opened connecting Harrisburg via Annville to Reading and built the first station in Annville in the late 1850s. It was acquired by the Philadelphia & Reading Railroad on March 20, 1858. The May 1870 schedule from the June 1870 "Travelers Official Guide of the Railways" showed the Philadelphia & Reading Railroad operated two westbound passenger trains from Allentown via Reading to Harrisburg stopping at Annville at 12:11 p.m. and 7:37 p.m. plus one train from Reading stopping at Annville at 8:38 a.m. for Harrisburg. Eastbound there were three passenger trains from Harrisburg to Allentown stopping at Annville at 9:01 a.m., 12:07 p.m., and 3:36 p.m. In addition, one train operated from Harrisburg only to Reading stopping at Annville at 5:01 p.m. On October 30, 1891, the Reading Railroad constructed the current Annville train station on Railroad Street east of the Lebanon Valley College. The Annville station was used for passenger service until 1963 and for freight service until 1976. Ron Rabena purchased the Annville Station from the Reading Railroad, completely dismantled it, and moved it to his New Cumberland area property. Rabena rebuilt the station, shortened its walls by 3 to 4 feet, and made the Buck Ridge Model Railroad Station to house and display his model trains. In 1988, Rabena sold the station to the Friends of Old Annville (FOOA). Blanche Schaeffer and Kathy Moe led the FOOA effort to purchase, move, and reconstruct the Annville Train Station at 155 N. Moyer Street in Annville on track-side land owned by Eagle Graphics Incorporated. FOOA, with a grant from the Pennsylvania Department of Community and Economic Development plus donations from local citizens, rebuilt and restored the station to its original height and original appearance.

On September 3, 2022, NS locomotive No. 1844 (originally NS No. 2549 EMD SD70 built in September 1994 and rebuilt into an SD70ACC by Progress Rail-Muncie, Indiana, in December 2019) is leading an eastbound intermodal train past the Annville Station in Annville, Pennsylvania. (*Brian A. Keates photograph*)

An NS westbound intermodal freight train is passing through Annville, Pennsylvania, on September 3, 2022 headed by 4,500-horsepower NS locomotive No. 1219 (EMD SD70IAC built in November 2019). The sixty SD70IAC units (Nos. 1175–1234) are equipped with IAC (Individual Axle Control) with one inverter per axle for better overall control of tractive effort which is the reason that these units were designated SD70IAC. (*Brian A. Keates photograph*)

On September 3, 2022, NS locomotive No. 4495 (originally NS No. 9142 GE Dash 9-44CW built in February 1998 rebuilt to type AC44C6M at NS Juniata Locomotive Shop in June 2021) is powering a westbound intermodal train through Annville, Pennsylvania, with the third locomotive being a trailing Union Pacific locomotive. (*Brian A. Keates photograph*)

10

Macungie Rail Fan Pavilion

Macungie, a borough in Lehigh County, Pennsylvania, that is a suburb of Allentown, has a railfan pavilion located on the 37-mile Reading Line of the Norfolk Southern Railway that runs from the Harrisburg Line at Wyomissing Junction in Wyomissing, Pennsylvania, to a junction with the Lehigh Line in Bethlehem, Pennsylvania. The open-air pavilion, located on the east side of the Highway Route 100 grade crossing on the south side of the tracks, has a parking lot with a covered viewing platform. The Reading & Lehigh Railroad, chartered on March 9, 1856, became the East Pennsylvania Railroad in April 1857. It completed a line between Reading and Allentown on May 11, 1859. This line became a through route between Harrisburg and New York City. The Philadelphia & Reading Railroad leased this line in 1869. After the Panic of 1893, the Reading Company was created to serve as a holding company for the Philadelphia & Reading Railway and the Philadelphia & Reading Coal & Iron Company. As a result of the Reading's final bankruptcy, it was merged into Conrail in 1976. The line through Macungie is now part of the Norfolk Southern Railway's Reading Line. Trains blow their horn for this crossing which adds to the enjoyment as they pass the pavilion. From Interstate 78, take exit 49A (to get on Route 100 South) and drive south to Macungie. Cross the double track (Norfolk Southern line) and the park is on the left.

NS locomotive No. 9727 (GE D9-44CW placed in service on December 1, 2001) is leading an eastbound intermodal train passing the Macungie, Pennsylvania, train viewing area located on Route 100 (Main Street) at Mill Street on November 19, 2022. This is also known as the Macungie Flower Park attracting both train watchers and flower enthusiasts. (*Brian A. Keates photograph*)

On November 19, 2022, NS locomotive No. 7005 (originally Chicago Northwestern Railway [CNW] 3,800-horsepower No. 8049 EMD SD60 built in September 1986, later Union Pacific Railroad [UP] No. 5966, later Helm Financial Corporation [HLCX] No. 5966 which leases rail cars and locomotives, still later NS No. 6547, and rebuilt to NS 4,000-horsepower No. 7005 [SD60E built in September 1986]) is powering an eastbound autorack train past the Macungie, Pennsylvania, train viewing area. (*Brian A. Keates photograph*)

NS locomotive No. 7652 (GE ES44DC placed in service on May 8, 2007) is leading a westbound intermodal freight train through Macungie, Pennsylvania, on November 19, 2022. (*Brian A. Keates photograph*)

11

Wernersville Train Station and Adjacent Area

Construction of the existing (granite and limestone with a slate roof) train station located on 20 Penn Avenue in Wernersville, a borough in Berks County, Pennsylvania (replacing the original small frame station that opened in 1881 on the Lebanon Valley Branch of the Philadelphia & Reading Railroad), began in 1926 and opened on June 6, 1927. The October 28, 1962 schedule shown in the *Official Guide of the Railways, December 1962,* shows there was one remaining daily passenger train in each direction connecting New York City via Allentown, Reading, Wernersville, Lebanon, and Hershey with Harrisburg. Railroad passenger service on the Reading's Lebanon Valley Branch ended in June 1963. The Heidelberg Heritage Society restored the Wernersville Passenger Station in the interests of historic preservation with the project completed in June 2013. The station is open for special events.

East of Wernersville via Highway 422 (Penn Avenue), the railroad is south of the highway and to photograph the railroad go south on Woodrow Avenue passing local roads (Ruth Street and Vestor Place where on street parking is available) to the Norfolk Southern Railway crossing of Woodrow Avenue in Sinking Spring, Pennsylvania, which is just east of the NS Wyomissing Junction. Photography should only be done from a public area.

Continuing east on Business Highway 422 (Penn Avenue) to Wyomissing, the railroad track becomes located on the north side of the highway. In Wyomissing, Pennsylvania, there is business parking on the north side of Penn Avenue in the area between Wyoming Avenue and Forest Avenue. Before parking at any private business parking lot, ask permission, support that local business, and never trespass on railroad property.

On September 18, 2022, NS locomotive No. 7670 (GE ES44DC placed in service on September 30, 2007) is leading an eastbound intermodal train past the historic Wernersville, Pennsylvania, passenger station. Years ago, city residents (largely from Philadelphia and New York City) traveled to Wernersville via the Reading Railroad to relax and enjoy the clean mountain air. (*Brian A. Keates photograph*)

The restored Wernersville, Pennsylvania (former Reading Railroad train station), is the location on September 18, 2022 of a westbound trash train powered by NS locomotive No. 8174 (GE ES44AC placed in service on April 27, 2017). The culmination of a thirteen-year restoration project of the Heidelberg Heritage Society, this station was rededicated on June 8, 2013. *(Brian A. Keates photograph)*

On November 14, 2021, NS locomotive No. 8033 (GE ES44AC placed in service on November 22, 2010) is powering a westbound intermodal train crossing Woodrow Avenue in the borough of Sinking Spring in Berks County, Pennsylvania. Sinking Spring was named for a spring located in the center of town where the water in this spring would sink into the ground occasionally giving the perception that it had disappeared. *(Brian A. Keates photograph)*

NS locomotive No. 9700 (GE D9-44CW placed in service on December 1, 2001) leads an NS westbound general manifest train at Woodrow Avenue in Sinking Spring, Pennsylvania, on a cloudy bright July 4, 2019. (*Brian A. Keates photograph*)

On September 2, 2019, NS locomotive No. 7567 (GE ES44DC placed in service on June 15, 2006) is heading a westbound general manifest through Sinking Spring, Pennsylvania. (*Brian A. Keates photograph*)

On June 27, 2021, an eastbound general manifest freight train headed by NS locomotive No. 4256 (originally NS No. 9025 GE Dash 9-44CW built in February 1997, rebuilt to type AC44C6M at NS Juniata Locomotive Shop in July 2019) is eastbound through the borough of Wyomissing in Berks County, Pennsylvania. (*Brian A. Keates photograph*)

On a cloudy bright May 27, 2019, NS locomotive No. 9536 (GE D9-44CW placed in service on December 21, 2000) is leading a westbound intermodal train through Wyomissing, Pennsylvania. (*Brian A. Keates photograph*)

12
Valley Forge Station

The Philadelphia & Reading Railroad began regular passenger service from Philadelphia via Port Kennedy (in Valley Forge, Pennsylvania) to Reading on December 5, 1839. The Pottsville via Reading to Philadelphia train service began on January 1, 1842. Constructed in 1911, the elegant Valley Forge passenger station welcomed visitors arriving to the Valley Forge State Park in Pennsylvania by train. Valley Forge was the location of the 1777–1778 winter encampment of the Continental Army under General George Washington during the American Revolutionary War. About 2 miles away was the Port Kennedy Station which in 2022 is in disrepair. The October 28, 1962 Reading Railway passenger schedule (used for this paragraph) showed one Philadelphia to Pottsville local train stopping at Port Kennedy at 7:36 a.m. An afternoon Philadelphia to Reading train stopped at Port Kennedy at 6:21 p.m. and Valley Forge at 6:25 p.m. A morning Reading to Philadelphia train stopped at Valley Forge at 6:55 a.m. and Port Kennedy at 6:58 a.m., and an afternoon Pottsville to Philadelphia local train stopped at Valley Forge at 5:05 p.m. and Port Kennedy at 5:11p.m. On Saturdays, only one morning Philadelphia to Pottsville train stopped at Port Kennedy at 7:36 a.m. On Sundays, no trains stopped at Port Kennedy and Valley Forge. The last day of passenger service on the Philadelphia via Reading to Pottsville line was June 30, 1981.

The station is near General George Washington's headquarters and the area his 12,000 troops plus 400 women and children troops arrived on December 19, 1777. Despite hardships, Valley Forge strengthened the U.S. Army with the development of basic training and military lessons provided by Prussian officer Baron von Steuben plus assistance from France was vital to secure victory. After the army left in June 1778, farmers returned and rebuilt the devastated area.

While the railroad line no longer has passenger service, the National Park Service uses the passenger station as an exhibit area. The nearby tracks are used by Norfolk Southern freight trains.

On December 29, 2022, NS locomotive No. 4432 (originally NS No. 9224 GE Dash 9-44CW built in May 1998, rebuilt to type AC44C6M at Wabtec/GE-Ft. Worth, TX, in March 2021) leads a westbound intermodal train past the Valley Forge Train Station located near Valley Forge Road (PA Route 23) and River Road at Valley Forge Park, Pennsylvania. (*Brian A. Keates photograph*)

The former Valley Forge Park Station (also known as Port Kennedy Station) located on the south side of the Schuylkill River on Station Lane north of Highway 23 is the location on July 2, 2022 of NS locomotive No. 3620 (GE ET44AC placed in service on June 1, 2016) powering an eastbound general manifest train heading for the NS Abrams Yard. (*Brian A. Keates photograph*)

On July 2, 2022, NS locomotive No. 8142 (GE ES44AC placed in service on September 7, 2013) is leading a westbound intermodal freight train under U.S. Highway Route 422 Bridge with the Schuylkill River (also under the Route 422 Highway Bridge) just north of and roughly parallel to the railroad. The former Valley Forge Park Station (also known as Port Kennedy Station) is slightly west of this point. (*Brian A. Keates photograph*)

13

East Penn Railroad

In 2007, East Penn Railways and Penn Eastern Rail Lines merged to form East Penn Railroad (EPR) which is owned by Regional Rail. EPR operates a number of mostly unconnected lines in Pennsylvania and Delaware. With the exception of two industrial switching operations, these were former Pennsylvania Railroad or Reading lines abandoned or sold by Conrail or its predecessors. Since that merger, the railroad has improved service and condition of lines having potential for customer growth and abandoning or selling weaker lines. EPR operates the following lines:

Bristol

The Bristol line has industrial trackage in the Bristol, Pennsylvania, area and interchanges with Conrail Shared Assets Operations (CSAO) which serves as a contract local carrier and switching company for its owners, CSX Transportation and Norfolk Southern Railway.

Lancaster Northern

The EPR line runs from Ephrata, Pennsylvania, northeast to interchange with Norfolk Southern Railway (NS) in Reading, Pennsylvania, using trackage rights on NS between Sinking Spring, Pennsylvania, and Reading.

Manheim

The line uses owned trackage in the Manheim, Pennsylvania, area.

North East Philadelphia

The line consists of owned trackage in Northeast Philadelphia, Pennsylvania, interchanging with CSAO in Bustleton.

Octararo

The line consists of owned trackage of the former Octararo Branch east to Chadds Ford where it connects to the Wilmington & Northern line. Southeastern Pennsylvania Transportation Authority (SEPTA) owns passenger rights on the Octararo line.

Perkiomen

The line, owned and operated by EPR, runs from Pennsburg, Pennsylvania, north to interchange with NS in Emmaus, Pennsylvania.

Quakertown

This line operates on the SEPTA-owned Bethlehem line from Quakertown, Pennsylvania, south to interchange with the Pennsylvania Northeastern Railroad in Telford, Pennsylvania.

Wilmington & Northern

The line, owned and operated by EPR runs from the interchange with CSX Transportation in Elsmere Junction, Delaware north to interchange with NS in Coatesville, Pennsylvania. In addition, the Octararo line connects with the Wilmington & Northern line in Chadds Ford, Pennsylvania.

York Industrial Track

The EPR line operates from interchange with NS in York, Pennsylvania, to Stony Brook, Pennsylvania.

Former Line: Colebrookdale

On September 6,1869, the Colebrookdale Railroad opened connecting Boyertown with Pottstown. It was leased by the Reading Railroad until 1976 when it became part of Conrail which planned to abandon the line. The PA Department of Transportation acquired the line and hired operators. The line was sold to Penn Eastern Rail Lines and was abandoned by East Penn Railroad (the successor to Penn Eastern Rail Lines). The Colebrookdale Railroad Preservation Trust was created in 2011, and tourist rail service began on October 18, 2014.

Former Line: Chester Valley

The line operated from King of Prussia, Pennsylvania, north to interchange with NS in Bridgeport, Pennsylvania. EPR abandoned the line in 2010, and tracks were dismantled in May 2011.

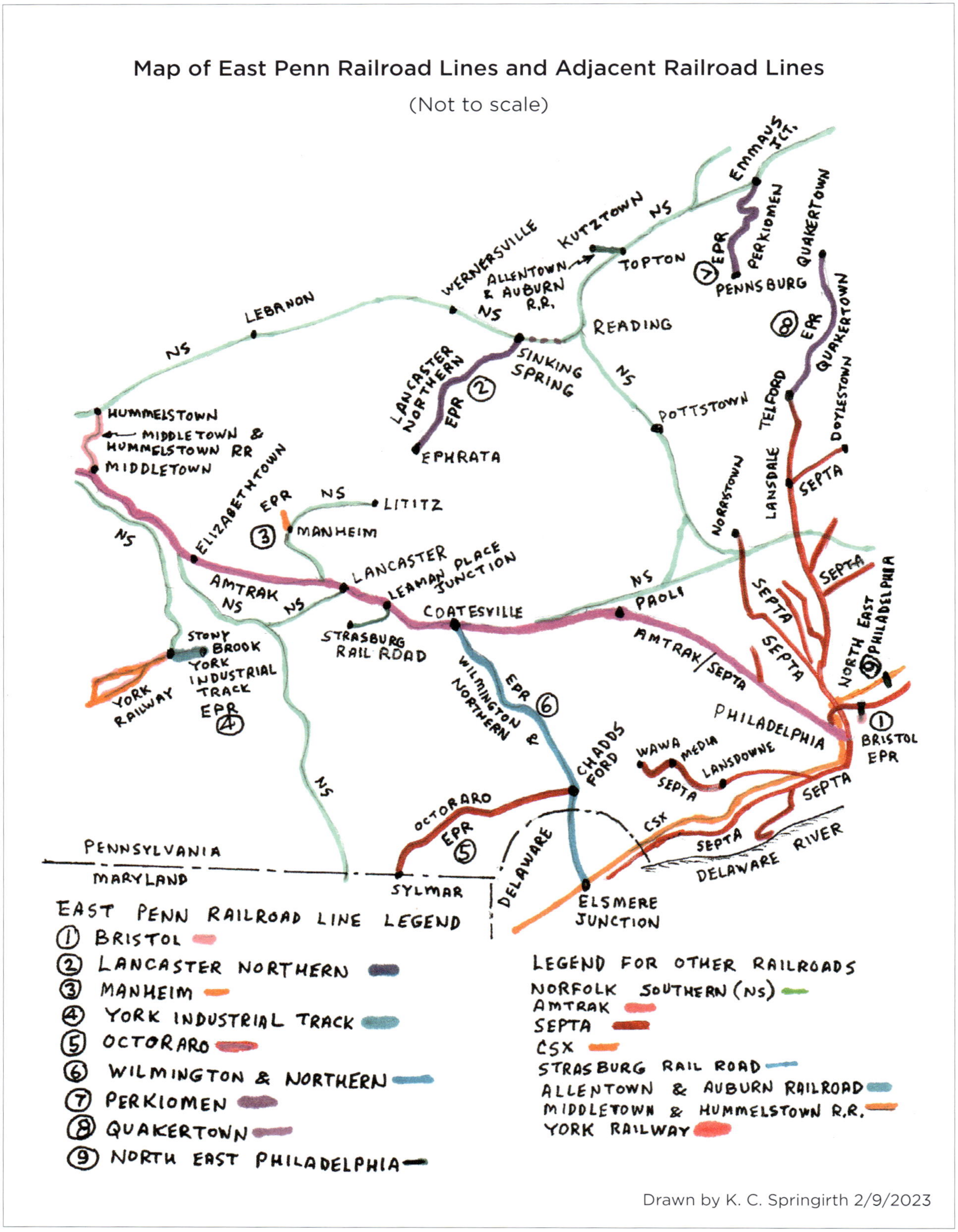

Norfolk Southern Railway interchanges with 261 short-line railroads including the East Penn Railroad and several other railroads in the southeastern region of Pennsylvania as shown in the above map. Short-line railroads help maintain rail service in places that might not have it which benefits customers and the economy.

On May 6, 2017, East Penn Railroad locomotives No. 1802 (reporting mark ESPN four-axle 1,800-horsepower GP 18 originally New York, Susquehanna & Western Railway No. 1802 built in August 1962) and No. 2801 (GP38 originally Chesapeake & Ohio Railway built in November 1967 and rebuilt as GP38-3) are waiting for the next assignment at the railroad's diesel facility in Reinholds an unincorporated community in West Cocalico Township in Lancaster County, Pennsylvania, Pennsylvania with PA Highway 897 crossing the tracks behind the locomotives. (*Brian A. Keates photograph*)

On September 8, 2020, East Penn Railroad No. 2801 locomotive is eastbound on the NS mainline at Wyomissing, Pennsylvania, where it has trackage rights from Sinking Spring to Reading. The reporting mark GMTX (just below the cab window) is part of Rail North America which leases locomotives to class 1 railroads, regional railroads, short line railroads, and industrial users. (*Brian A. Keates photograph*)

On January 3, 2023, East Penn locomotive No. 2801 is coming off the Epherta Branch of the East Penn Railroad to get on the Harrisburg Line at Sinking Spring, Pennsylvania. When the Reading Company became part of Consolidated Rail Corporation on April 1, 1976, the Harrisburg Line was formed from the original Reading Company main line from Philadelphia to Reading and the original Lebanon Valley Railroad line from Reading to Harrisburg. The Harrisburg line became part of Norfolk Southern Railway on June 1, 1999. (*Brian A. Keates photograph*)

NS locomotive No. 9899 (GE D9-44CW placed in service on September 26, 2004) is leading an eastbound intermodal train past East Penn Railroad locomotive No. 2801 handling a local run at Sinking Spring, Pennsylvania, on January 3, 2023. (*Brian A. Keates photograph*)

14

R.J. Corman Railroad/Pennsylvania Lines

The R.J. Corman Railroad/Pennsylvania lines (reporting mark RJCP) are lines formerly owned by the Pennsylvania Railroad (earlier by the Cambria & Clearfield Railway), New York Central (earlier by the Beech Creek Railroad plus Beech Creek Extension Railroad subsidiaries), and the Cherry Tree & Dixon Railroad. These lines came under common control with the acquisition of the New York Central Railroad by the Pennsylvania Railroad forming Pennsylvania New York Central Transportation Company (shortened to Penn Central) on February 1, 1968, which passed to Consolidated Rail Corporation (Conrail) on April 1,1976. Conrail sold these lines to RJCP in 1995.

With its nicely painted locomotives, R.J. Corman looks more like a large class 1 railroad than a moderately sized short-line railroad. Its founder, Richard J. "Rick" Corman began his career in railroad construction with a goal to become one of the United States leading railroad service providers. On November 14, 2022, the R.J. Corman Company purchased its nineteenth railroad, creating Raleigh & Fayetteville, a 43-mile branch line, and leased a 20-mile branch line from Norfolk Southern Railway.

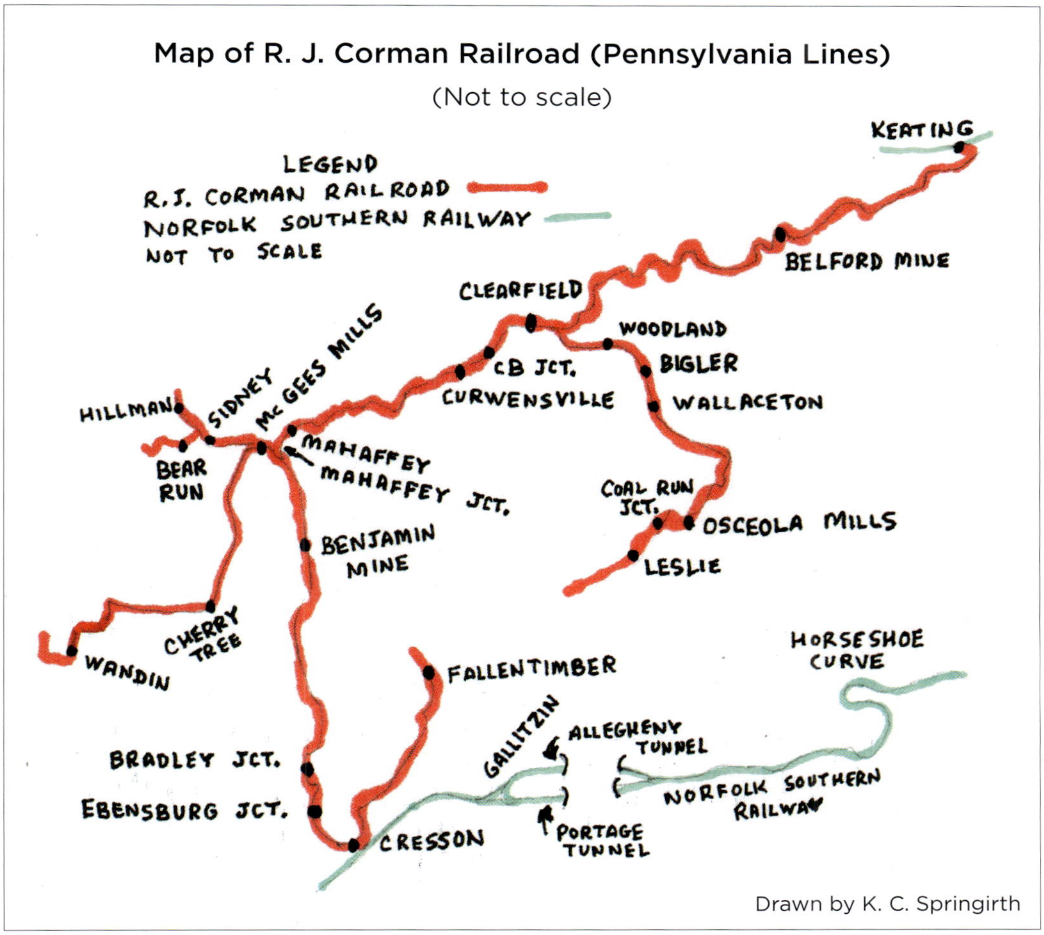

On the above map, the R.J. Corman Railroad, interchanging with the Norfolk Southern Railway at Cresson and Keating, operates 243.42 miles of track in Cambria, Centre, Clearfield, Clinton, and Indiana Counties in Pennsylvania.

On August 22, 2018, R.J. Corman Railroad/Pennsylvania lines (RJCP) locomotive No. 7123 (EMD SD40-2 originally built in December 1978 as Burlington Northern Santa Fe Railway No. 7123) is on a siding next to the Cresson Steel Company in Cresson, Pennsylvania. The railroad's founder Richard J. Corman took pride in the RJCP locomotive fleet, which is why almost forty years after the locomotive was built (the locomotive in its bright red paint scheme) it looks new, like it just came out of the paint shop. *(Brian A. Keates photograph)*

RJCP locomotive No. 7109 (EMD SD40-2 originally built in December 1978 as Burlington Northern Santa Fe Railway No. 7109) and trailing RJCP locomotive No. 7095 (EMD SD40-2 built in November 1978 as Burlington Northern Santa Fe Railway No. 7095) are leading a grain train northbound just outside of Cresson, Pennsylvania, approaching a local farm crossing on October 5, 2022. RJCP locomotives are painted red because Corman's first pickup truck was red and he said, "You're better off if you don't change a lot of stuff that seems to be working." *(Brian A. Keates photograph)*

New York, Susquehanna & Western Railway

In 1881, the New Jersey Midland Railroad was consolidated with five other railroads to form the New York, Susquehanna & Western Railway (NYS&W). The NYS&W fell on hard times during the 1957 recession, and it lost its western connection to the Lehigh & New England Railroad (L&NE) when the L&NE ended operation in 1961. In 1980, the Delaware Otsego Corporation purchased the NYS&W. In 2023, the NYS&W operates over 400 miles of track in New York, New Jersey, and Pennsylvania.

Map of New York, Susquehanna and Western Railway

(Not to scale)

Drawn by K. C. Springirth

The above map shows the NYS&W operation in New York, New Jersey, and Pennsylvania. It transports feed ingredients, building materials, chemicals, and aggregates for New York state customers, and plastics, lumber, food products, paper products, motor vehicles, chemicals, aggregates, and metals for New Jersey and Pennsylvania customers.

Waiting for the next assignment near Binghamton, New York, on July 30, 2022, is New York, Susquehanna & Western Railway (NYSW) locomotive No. 3012 (EMD SD40T-2 originally No. 8242 built for Southern Pacific Railroad in April 1980). This locomotive is known as a tunnel motor with low mounted air intakes and larger radiators which solved the problem of a locomotive overheating which occurred inside the long tunnels and snow sheds on the Southern Pacific Railroad mainline in the Sierra Nevada. (*Brian A. Keates photograph*)

On July 8, 2017, NYSW locomotive No. 3018 (EMD SD40-2 originally No. 3267 built for the Southern Pacific Railroad in January 1978) is shifting freight cars on NYSW trackage near Binghamton, New York. (*Brian A. Keates photograph*)

16

Juniata Valley Railroad

The Sunbury & Lewistown Railroad opened a 43.5-mile line from Lewistown to Selinsgrove on December 1, 1871. With little traffic on the rural line, it was sold at a sheriff's sale in May 1874, reorganized as the Sunbury & Lewistown Railway on February 1, 1876, and leased by the Pennsylvania Railroad (PRR). Under PRR control, it became an important shortcut for moving anthracite coal from Wilkes Barre westward avoiding Harrisburg, and for moving perishables to New York markets via the Lehigh Valley Railroad at Mt. Carmel, avoiding both Harrisburg and Philadelphia. With industrial decline during the 1950s, the middle of the line was abandoned beginning in 1957. Beginning April 1, 1976, Conrail operated railroad lines that served Lewistown. The Susquehanna Economic Development Authority Council of Governments Joint Rail Authority (SEDA-COG JRA) was formed in 1983 to preserve essential rail service and to further economic development and job creation through improvement and expansion of rail infrastructure. The North Shore Railroad Company and Affiliates was established in 1984 and is composed of six short-line railroads operating on 256.2 miles of track in Central Pennsylvania. One of the six railroads is the Juniata Valley Railroad line that began operating August 19, 1996 on 18.5 miles of former Conrail track from Lewistown to Burnham, Maitland, and Mifflin County Industrial Development Corporation Park (now owned by SEDA-COG JRA) interchanging with the Norfolk Southern Railway at Lewistown, Pennsylvania.

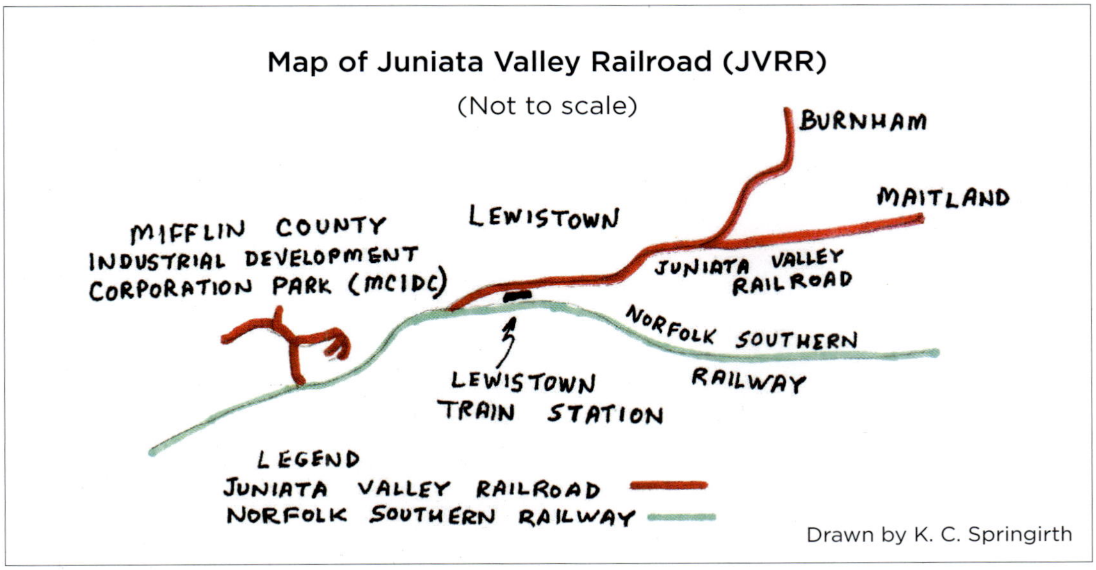

The Juniata Valley Railroad (JVRR) map shows its interchange points with the Norfolk & Western Railway west of the Lewistown station and further west to the Mifflin County Industrial Development Corporation Park (MCIDC).

On July 20, 2022, Juniata Valley Railroad (JVRR) diesel switcher locomotive No. 2106 (1,200-horsepower EMD SW9 originally built in November 1993 for the Pittsburg [Pittsburgh was spelled Pittsburg because the United States Board of Geographic Names advocated spelling Pittsburgh without the h from 1891 to 1911] & Shawmut Railroad as No. 236 and later renumbered 1865) is ready to switch some cars at Lewistown, Pennsylvania. The keystone-shaped number plate pays tribute to this line's Pennsylvania Railroad heritage. (*Kenneth C. Springirth photograph*)

At the Lewistown, Pennsylvania, yard, JVRR locomotive No. 2106 is switching cars on July 20, 2022. At the time this picture was taken, this locomotive, painted Tuscan red with yellow trim, was sixty-nine years old and looked like it was brand new. Commodities handled by this railroad include scrap, finished metals, plastics, fertilizer, and pulp. (*Kenneth C. Springirth photograph*)

17

Buffalo Southern Railroad

Incorporated on March 23, 1872, the Buffalo & Jamestown Railroad (in New York State) was built with the aid of $1 million from the city of Buffalo and another $1 million was donated by the towns along the line. On October 20, 1874, the line was completed in New York state from Buffalo to Gowanda, and was completed to Jamestown, New York, in July 1875. Never financially successful, the line was acquired by the Buffalo & Southwestern Railroad (by a December 12, 1877 deed) which was leased to the New York, Lake Erie & Western (Erie Railroad) on August 1, 1880. The Erie Railroad merged the railroad by owning 100 percent of the stock on November 19,1895. Passenger service ended on the Erie Railroad's Jamestown–Buffalo line on June 4,1950.

The Buffalo Southern Railroad (reporting mark BSOR) operates the 32 miles of track (owned by Erie County, New York and leased from the Erie County Industrial Development Agency line) from Buffalo, New York, to the Erie County line at Gowanda, New York, where it interchanges with the New York and Lake Erie Railroad. BSOR (by its mostly Alco fleet of locomotives) transports fertilizers, lumber, scrap metal, animal feed, and grains. BSOR hosts the Buffalo, Cattaraugus & Jamestown Scenic Railway, a non-profit organization operating diesel- and steam-powered excursions in the Hamburg, New York, area.

Map of Buffalo Southern Railroad

(Not to scale)

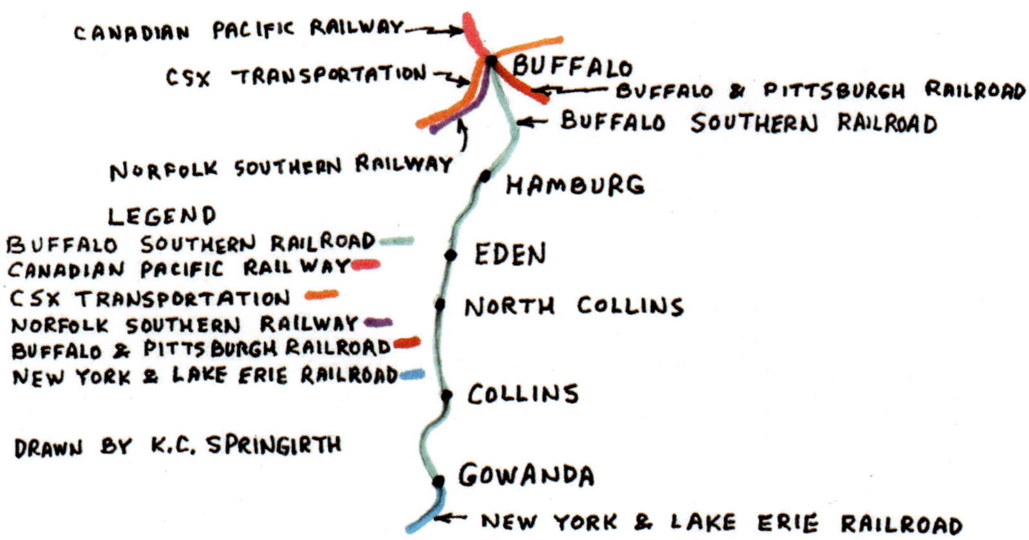

Drawn by K. C. Springirth

The above map shows the Buffalo Southern Railroad connecting Buffalo via Hamburg, Eden, and North Collins to Gowanda, New York.

On September 22, 2012, Buffalo Southern Railroad 1,000-horsepower diesel electric switcher locomotive No. 93 (Alco S-4 originally built in August 1951 as South Buffalo Railway No. 93) is powering an excursion train at Eden Center, New York. (*Kenneth C. Springirth photograph*)

Buffalo Southern Railroad 1,800-horsepower diesel electric locomotive No. 5010 (originally Delaware & Hudson Railway No. 5010 Alco RS-11 built in May 1961) on October 1, 2022 is powering a Buffalo, Cattaraugus & Jamestown Scenic Railway Company (BC&JSRC) excursion train operating on the Buffalo Southern Railroad at Hamburg, New York. The mission of the BC&JSRC is to encourage public participation in passenger train rides to learn about railroads: "in a manner that is safe, educational, enjoyable, economically self- sustaining, and a credit to the legions of man and women who have worked for or used the railroad before us and will in the future." (*Kenneth C. Springirth photograph*)

18

Western New York & Pennsylvania Railroad

The Atlantic & Great Western Railroad (A&GW) completed a line west from Salmanca, New York, to Meadville, PA, on November 10, 1862. The A&GW became the main line of the Erie Railroad to Chicago. The Interstate Commerce Commission (ICC) approved the merger of the Erie Railroad and Delaware, Lackawanna & Western Railroad into the Erie Lackawanna Railroad on October 17, 1960, which was merged into the Consolidated Rail Corporation (Conrail) on April 1, 1976. On June 1, 1999, Norfolk Southern Railway (NS) acquired 58 percent of Conrail's assets—roughly 6,000 route miles—which included the Corry, PA to Hornell, NY, line, and CSX acquired 42 percent of Conrail's assets—roughly 3,600 miles. On April 9, 2001, the Western New York & Pennsylvania Railroad (WNYP) began a long-term lease of the NS line from Corry, Pennsylvania, to Hornell, New York. In January 2002, the WNYP acquired the Corry-Meadville line from the Northwest Pennsylvania Rail Authority. NS coal trains began using the Meadville–Hornell line in 2004. In December 2005, the WNY&P leased from NS the former Erie RR from Meadville via Oil City to Rouseville, Pennsylvania. NS coal trains that used the Meadville–Hornell line were suspended on January 1, 2012.

Map of Western New York & Pennsylvania Railroad

(Not to scale)

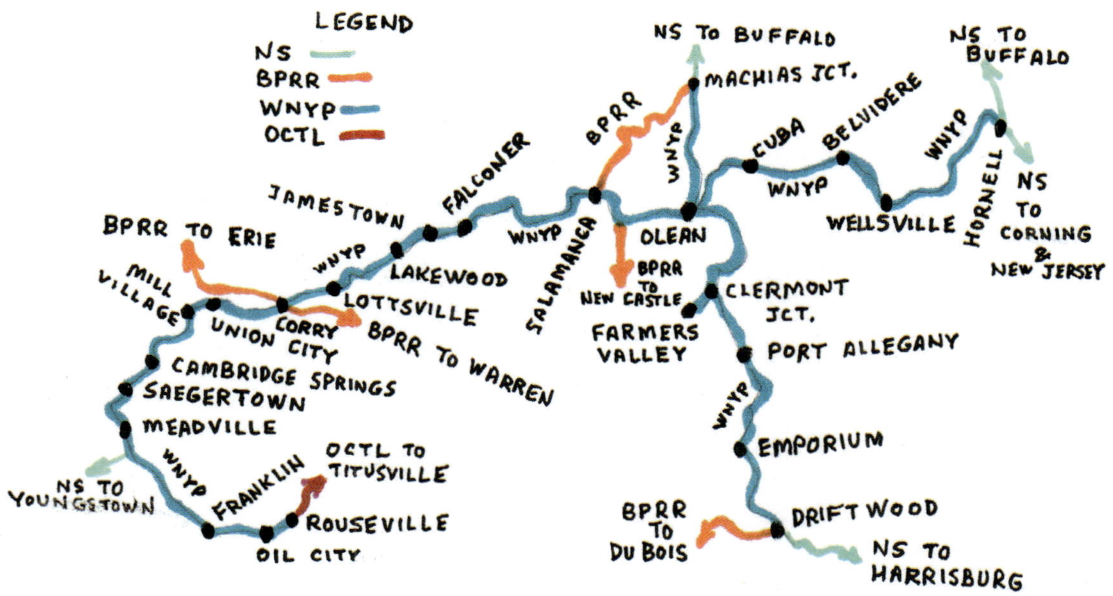

Drawn by K. C. Springirth

As shown from the above map, the 330-mile Western New York and Pennsylvania Railroad extends across southwestern New York and northwestern Pennsylvania from Hornell, New York, to Meadville and Oil City, Pennsylvania, and north to south from Machias Junction, New York, via Olean, New York, to Driftwood, Pennsylvania. The Oil Creek & Titusville Railroad (OCTL) acquired the line from Titusville to Rynd Farm from Conrail and began tourist train operations on July 18, 1986; freight operations on September 25, 1986. OCTL interchanges with WNYP. WNYP interchanges with Norfolk Southern (NS) and the Buffalo & Pittsburgh Railroad (BPRR) at various locations.

On June 30, 2011, Western New York & Pennsylvania Railroad (WNYP) six-axle 3,600-horsepower locomotive No. 637 (Montreal Locomotive Works type M636 originally built as Canadian National Railway No. 2312 in April 1970) is at the WNYP locomotive shop, yard, and office area at Olean, New York. Other railroads this locomotive operated on include Quebec Cartier Railroad No. 47 and New York, Susquehanna & Western Railway No. 47. (*Kenneth C. Springirth photograph*)

WNYP Railroad locomotive No. 4264 four-axle 2,500-horsepower (Alco C425 originally built as Spokane Portland & Seattle Railway No. 326 in June 1966) is waiting for the next assignment at the WNYP Olean, New York, yard on June 30, 2011. This locomotive has also operated on other railroads including Burlington Northern Railroad No. 4264, Arkansas & Missouri Railroad No. 4264, Morristown & Erie Railroad No. 4264, and Massachusetts Central Railroad No. 4264. (*Kenneth C. Springirth photograph*)

At the former Erie Railroad passenger station in Jamestown, New York, on August 6, 2011, two former Pennsylvania Railroad EMD type E8A locomotives (No. 5711 built in October 1952 and No. 5809 built in January 1951) are on a special passenger train excursion. Both of these locomotives were extensively restored and painted in the Pennsylvania Railroad wide stripe paint scheme by the Juniata Terminal Company at its Philadelphia, Pennsylvania, facility. (*Kenneth C. Springirth photograph*)

On August 7, 2011, a special passenger excursion on the WNYP Railroad (passing the former Erie Railroad passenger station in the borough of Utica, Venango County, Pennsylvania) is headed by WNYP four-axle 3,000-horsepower locomotive No. 430 (Alco type C430 built in November 1967). This locomotive has also been used on these railroads: Penn Central No. 2050, Conrail No. 2050, and New York, Susquehanna & Western Railway No. 3000. (*Kenneth C. Springirth photograph*)

With the Smock Memorial Bridge in the background, WYNP four-axle 2,400-horsepower locomotive No. 426 (Alco type C424 originally Belt Railway of Chicago No. 603 built in May 1966) is on the siding at the rail yard in Meadville, Crawford County, Pennsylvania, on August 2, 2011. This yard is an interchange point between Norfolk Southern Railway (NS) and WYNP. (*Kenneth C. Springirth photograph*)

On May 5, 2012, WNYP locomotive No. 430 is waiting for the next assignment at the Meadville, Pennsylvania, railroad yard close to the four-lane 1,613-foot-long Smock Memorial bridge (built in 1959) that carries highway Routes 6, 19, and 322 over French Creek and over tracks of NS and WNYP in Vernon and West Mead Townships and the City of Meadville. (*Kenneth C. Springirth photograph*)

WNYP locomotive Nos. 430 (Alco type C430 built in November 1967) and 435 (Alco type C424 originally Canadian Pacific Railway No. 4228 built in 1965) are switching cars near Meade Avenue in Meadville, Pennsylvania, on July 17, 2022. (*Kenneth C. Springirth photograph*)

On July 17, 2022, WNYP locomotives Nos. 435 and 430 have brought in 6,000-horsepower WNYP locomotive No. 6003 (GE type AC60CW originally built as CSX No. 5013 in April 2000) just south of Mead Avenue in Meadville, Pennsylvania, for viewing at the restored Meadville station site. The WNYP purchased eight 6,000-horsepower locomotives (Nos. 6000–6007) because of their ability to pull heavy loads. (*Kenneth C. Springirth photograph*)

Repainted WNYP locomotive No. 6003, one of eight GE type AC60CW locomotives, is on display for visitors at the refurbished Meadville Depot on July 17, 2022. (*Kenneth C. Springirth photograph*)

In this July 17, 2022 view, to the right of WNYP No. 6003 is the former Meadville Farm & Garden building that was purchased and renovated by members of the French Creek Valley Railroad Historical Society. The building was built by the Meadville Railway and the first train arrived in October 1881, reorganized as the Meadville & Linesville Railway in February 1884, became Meadville Conneaut Lake & Linesville Railway in June 1891, leased to the Pittsburgh Shenango & Lake Erie which became Pittsburgh Bessemer & Lake Erie on January 18, 1897, became Bessemer & Lake Erie Railroad on December 31, 1900, and the Meadville branch from Meadville Junction to the city of Meadville ended operation on May 13, 1977. (*Kenneth C. Springirth photograph*)

19

Westfield, New York's Amazing Rail Site

Over the years, an informal rail park was developed by a small group of friends who wanted a place to meet and watch trains, on North Gale Street between the double-track main line plus siding of CSX and the single-track Norfolk Southern Railway north of West Main Street (U.S. Highway 20) in Westfield, Chautauqua County, New York. Robert J. Augur and volunteers, without any government funding, cleared considerable space between the two railroad lines which improved the view for watching and photographing trains, enhanced railroad safety for train crews, and facilitated the work of the local police department in their daily patrolling of Westfield. North Gale Street passes through the informal park with the western area having a grassy area with picnic tables, benches, bulletin board (with notices of rail related events), and a parking area for automobiles. The eastern area of the park is more wooded with a nice area cleared of underbrush. Westfield's historic downtown is less than two miles east of the rail park on Main Street.

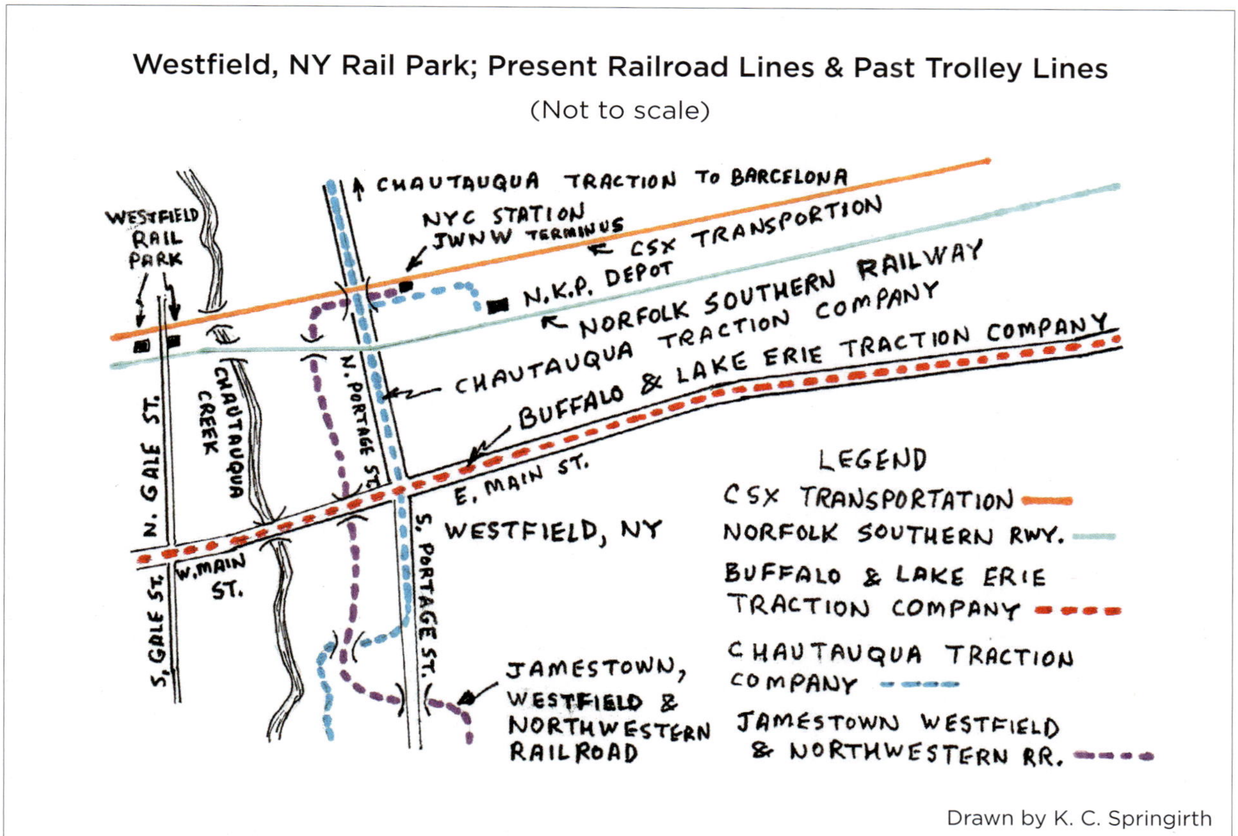

This map shows the informal North Gale Street rail park in Westfield, New York, between CSX (earlier New York Central Railroad) and Norfolk Southern Railway (earlier New York Chicago & St. Louis Railroad). Dotted lines show the former interurban trolley lines that once served Westfield, New York.

On October 12, 2022, NS locomotive No. 3335 (EMD SD40-2 originally Conrail No. 6365 built in June 1977 was transferred to NS on June 1, 1999) is switching cars at the Welch Grape Juice facility in Westfield, Chautauqua County, New York, the largest grape-growing county outside of California. (*Kenneth C. Springirth photograph*

An eastbound NS intermodal train on June 21, 2022 is crossing North Gale Street in Westfield, New York, powered by Canadian Pacific Railway GE type ES44AC locomotives No. 8948 (built in October 2011) and No. 8934 (built in September 2011). These locomotives were designed to meet the U.S. EPA's Tier 2 locomotive emissions that took place in 2005. (*Kenneth C. Springirth photograph*)

On June 21, 2022, westbound NS locomotive No. 9645 (GE D9-44CW placed in service on February 16, 2001) is passing over the North Gale Street crossing at Westfield, New York. South of the tracks is private property, and it is important for visitors at the rail park to keep off of private property. All it takes is one careless individual doing the wrong thing, and this wonderful viewing site could be lost. *(Kenneth C. Springirth photograph)*

Eastbound NS intermodal train on October 26, 2022, is headed by 4,400-horsepower Burlington Northern Santa Fe Railway locomotive No. 8380 (one of eighty-two GE type ES44C4 units [Nos. 8318–8399] built in 2015) and 6,000-horsepower Union Pacific Railroad No. 7522 (one of forty-five GE type AC6000CW units [Nos. 7510–7554] built in 1998). This picture was taken at the Gale Street crossing in Westfield, New York. *(Kenneth C. Springirth photograph)*

On October 13, 2022, three employees of Robert Vorst Paving Company of Ohio are repaving the crossing of two of the CSX tracks (paving at the northernmost CSX track was not replaced) at the Gale Street Crossing in Westfield, New York, which is north of the single-track NS Gale Street Crossing. It was an educational experience to watch this operation. A nearby CSX employee notified the paving crew in advance when the next train was coming. (*Kenneth C. Springirth photograph*)

The skilled paving crew does not waste any time in leveling the paving material at the Gale Street, Westfield, New York, CSX crossing on October 13, 2022, because the trains have the right of way. A CSX employee notified the paving crew in advance when a train was coming. (*Kenneth C. Springirth photograph*)

A westbound CSX general manifest train is passing by the Gale Street crossing repaving operation with 4,000-horsepower six-axle CSX Distributed Power Unit (DPU) No. 7509 (GE type C40-8 built in May 1989) in the center of the train on October 13, 2022. The railroad grade crossing repaving operation requires a lot of advance planning between the railroad and paving company. (*Kenneth C. Springirth photograph*)

An eastbound NS intermodal train is approaching the North Gale Street grade crossing in Westfield, New York, on October 25, 2022 powered by GE type AC44CW Union Pacific Railroad locomotives No. 6586 (built in April 1997) and No. 6512 (built in May 2000). To the right of the locomotives is the CSX signal tower just north of the nicely sized parking area of the Gale Street rail park. (*Kenneth C. Springirth photograph*)

On October 25, 2022, NS locomotives No. 1217 (EMD SD70IAC built in November 2018), No. 9493 (GE type D9-44CW built on April 29, 2000), and No. 6224 (NS model designation SD33ECO that was originally Baltimore & Ohio Railroad [BO] No. 7484 (EMD SD40 built in May 1967), second owner Central Railroad of New Jersey [CNJ] No. 3062, third owner Conrail [CR] No. 6287, and rebuilt as CR No. 6996 at Conrail's Juniata Shop to SD40-2 specifications becoming NS No. 3445, and was rebuilt and renumbered 6224) are powering an eastbound train passing by the North Gale Street rail park in Westfield, New York. The park provides a convenient location to photograph two mainline railroads. (*Kenneth C. Springirth photograph*)

At the rear of the above NS train is 3,000-horsepower NS locomotive No. 6224 (rebuilt to type SD33ECO from NS No. 3445 SD40-2 by NS in June 2021) painted in a special two-tone green, white, and black scheme. This locomotive, partially funded with federal and state grants, was assigned for use in the Buffalo, New York, area. (*Kenneth C. Springirth photograph*)